Ethnobotanical Wisdom of the Tribals in the Palni Hills

Ethnobotanical Wisdom of the Tribals in the Palni Hills

By

Rev. Dr. S.M.John Kennedy S.J.,

2008

DAYA PUBLISHING HOUSE

Delhi - 110 035

ISBN10 81-7035-554-0
ISBN13 978-81-7035-554-0

Published by : **Daya Publishing House**
1123/74, Deva Ram Park
Tri Nagar, Delhi - 110 035
Phone: 27383999
Fax: (011) 23260116
e-mail : dayabooks@vsnl.com
website : www.dayabooks.com

Showroom : 4760-61/23, Ansari Road, Darya Ganj,
New Delhi - 110 002
Phone: 23245578, 23244987

Laser Typesetting : **Classic Computer Services**
Delhi - 110 035

Printed at : **Chawla Offset Printers**
Delhi - 110 052

PRINTED IN INDIA

Dedicated to

My Father
Mr. S.MICHAEL
&
My Mother
Mrs. S.M. PAPPA
&
All the Tribals
in the PALNI HILLS

Acknowledgements

I am extremely grateful and thankful to the following persons and institutions for their help in the successful completion and publication of my research work:

1. To Rev. Dr. V.S. Manickam, my guide, whose persevering and strenuous efforts in guiding me are imprinted in each and every page of this thesis.
2. To Rev. Dr. Ignacimuthu, for his message and constant encouragement and support in all scientific works.
3. To PEAK team members especially to Frs. S.Arockiam, A.Kulandai who helped me through and through in the collection of data and complemented my data with their experiences in working among the tribals for the past fifteen years.
4. To all the tribal people in the Palni Hills without whom this research would have remained only a dream.
5. To Sacred Heart College in Kodaikanal for providing me the board and lodging facilities.
6. To my friends, superiors, mother, brother, sisters, relatives and companions for their constant source of help, support and encouragement.

7. To Mr. Anil Mittal of Daya Publishing House, New Delhi, for giving a beautiful shape to this book and publishing it.
8. Finally to the ALMIGHTY for everything.

Rev. Dr. S.M.John Kennedy S.J.,
St. Mary's Residence
P.B. No. 6, Dindigul – 624 001, T.N.

Foreword

Fourteen percent of the total 80 million sq. km. area of the land surface is covered with forests. Today throughout the world an awareness of the need for documentation of traditional knowledge of medicines of the tribals and village communities living in or close to the forests is emerging. In India the Foundation for Revitalization of Local Health Traditions in Bangalore has been mostly concentrating on this aspect along with conservation of medicinal plants. Throughout the world about 35 to 55 thousand plant species have been used by people for medicinal purposes. Currently India is exporting medicinal plants worth Rs. 800 crores annually but the tragedy is that most of the species are collected from nature. India's land area is 2 per cent of the world's land surface but contains 8 per cent of the global biodiversity. The Western Ghats, of India harbours 10 per cent of the Indian flora *i.e.* 1600 species of which 105 are medicinally useful. Forty two percent of the angiosperm species on the summits of the Western Ghats are endemic and hence this series of mountains is compared with Island as far as endemism is considered. In these Ghats 1500 angiosperm species are endemic *i.e.* 75 per cent of the endemic plants of peninsular India. The Palni

Hills, the eastward offshoot of the Western Ghats with an area of 2040 sq. km. has an angiosperm flora of more than 2500 species. Attempts are being made to declare the Palni Hills as a National Park.

The Ph.D. thesis of Rev. Dr. S.M.John Kennedy, entitled "Ethnobotanical Elements of the Tribals in the Palni Hills with Special Emphasis on the Traditional Medicinal Plants" was highly commended by both the foreign and Indian examiners for the award of the degree mainly because of the volume of data and with regard to the plants used by two tribal groups namely, Pulayans and Palliyans inhabiting various forest segments particularly in the Lower Palnis. The author has personally met the tribals and enquired them about the various ways and ailments for which they use the plants, particularly for their use of plants for medicinal purpose. The readers of this book would be amazed to know that the two tribal groups are using 326 plants for curing 192 diseases. Apart from the use of plants for medicinal purpose, the tribals use them for various other purposes like rituals performed for deities, for different kinds of festivals in their culture and also for food for themselves and for their domestic animals.

A remarkable feature of this book is that the author has presented copious data for each plant species such as the botanical name, vernacular name, habit, parts used, how used, diseases cured, report by a specific person, and the area where the plant is collected. The data presented in this regard are very relevant and useful for readers interested in ethnobotany of the Palni Hills.

Under the heading "Conclusions, Suggestions and Recommendations" the author deals with the economic conditions of the tribals, tribal health care, ethnomedicine, with positive and negative aspects, proposals for revival of ethnomedicine, conservation of medicinal plants, tribal development, towards a new forest policy, development of the tribals in the Palnis and finally common property resources and people's participation. All these aspects are of tremendous practical value for empowering the tribals in socio-economic conditions and also for conservation of plant species through people's participation.

This book on Ethnobotany will be useful not only for scientists of different kinds but also for college students who would like to take up this branch of biology for their researches.

V.S. Manickam

Rev. Dr.V.S.Manickam, S.J.
Director
Centre for Biodiversity and Biotechnology
St.Xavier's College
Palayamkottai – 627 002
Tamil Nadu

From the Author

Aborigines have the store house of the wealth of traditional wisdom. Modern generations, in spite the tremendous scientific advancements, need to learn a lot from this folk wisdom. Traditional wisdom is experience based and tested and interiorized after centuries of trial and error methods.

Tribals in the Palni hills are the earliest known inhabitants in these hills. I have been working among these people since 1992. My personal acquaintance with them made me realize the immense wealth of their traditional wisdom, the need to preserve them and the urgency of communicating that to the modern generations for a healthy and harmonious living. The outcome of such an endeavour is this book.

I hope, this book will take us back to our roots and facilitate a healthy blossoming of the present and the future.

Rev. Dr. S.M.John Kennedy S.J.,
St. Mary's Residence
P.B. No. 6, Dindigul – 624 001, T.N.
E-mail: kennedy-john@mailcity.com
jkkenny@gmail.com

Loyola College
Entomology Research Institute *Chennai*

India 600034
Tel. (Off): +91(44)28178348
(Res) : +91(44)28178210
Fax : +91(44)28174644
e-mail : eri_lc@hotmail.com

Rev. Fr. Dr. S. Ignacimuthu, s.j.
Director *Ph.D., D.Sc.,*

Message

Human beings have always been largely dependent on plant resources for their food, shelter, medicine and other needs. Living close to nature, the tribal communities have been having intimate contact with plants and have greatly benefited by them. This book on 'Ethnobotanical Wisdom of the Tribals in the Palni Hills' written by Rev. Dr. S.M. John Kennedy has documented every single plant used by the tribals living in the Palni Hills. He also gives a detailed description about the life and customs of tribals and their population size. He also vividly portrays the socio-economic status of the underprivileged tribal community at a time when traditional knowledge is getting eroded. It is really praise worthy that the author has taken great pains to document this knowledge before it gets lost with the passage of time. The author also pleads for safeguarding the heritage and culture of the tribals in the Palni Hills.

Apart from the in-depth study, the author also gives a number of very good suggestions for the policy makers and voluntary workers to improve the lot of tribal people. I would like to congratulate Rev.

Dr. John Kennedy for this good work. Let me also complement Mr. Anil Mittal of Daya Publishing House, New Delhi for bringing out this book in an attractive manner.

Rev. Dr. S. Ignacimuthu, s.j.

Former Vice Chancellor,
Bharathiar University,
Coimbatore and University of Madras,
Chennai

Contents

	Acknowledgements	*vii*
	Foreword	*ix*
	From the Author	*xiii*
	Message	*xv*
1.	Introduction	1
2.	Ethnobotany and its Scope	10
3.	The Palni Hills in the Western Ghats	24
4.	Sociological Aspects of Palliyans and Pulayans	48
5.	Cultural Aspects of the Palliyans and Pulayans	60
6.	Health Care and Medical Practices	76
7.	Ethnobotanical Wealth of the Tribals	96
8.	Alienation and Exploitation of the Tribals	196
9.	Tribal Development and Conservation of Forests	208
	Appendices	227
	Index	245

Chapter 1
Introduction

Human Dependence on Plants

Evolution of human life and culture has directly or indirectly been associated with and influenced by the surrounding environment. Primitive people live closely associated with nature and chiefly depend on it for their survival. Their dependence on plants around them made them acquire the knowledge of economic and medicinal properties of many plants by trial and error methods. Consequently they became the store-house of knowledge of many useful as well as harmful plants, accumulated and enriched through generations and passed on from one generation to another, without any written document.

World wide, tens of thousands of species of higher plants and several hundred lower plants are currently used by human beings for a wide variety of purposes such as food, fuel, fibre, oil, herbs, spices, industrial crops and as forage and fodder for domesticated animals.[1]

Many people, especially in poorer countries rely on these wild plants for food, construction materials, fuel wood, medicine and many other purposes. Traditionally local communities worldwide are extremely knowledgeable about plant and other natural resources, on which they are so immediately and intimately dependent. Unfortunately, much of this wealth of knowledge is today becoming lost as traditional cultures become eroded. Ethnobotanists can play very useful roles in rescuing this disappearing knowledge and returning it to local communities. In this way local ethnobotanical knowledge can be conserved as part of living cultural-ecological systems, helping to maintain a sense of pride in local cultural knowledge and practice and reinforcing links between communities and the environment, so essential for conservation.[2]

It is therefore, important that before this rich unwritten folk-lore on uses of plants and plant resources is lost forever through the recent accelerated 'civilization' of the aborigines (tribals), it must be properly documented and preserved.[3]

Tribals in India

Eminent ethnologists viz: Risley, Grigson, Schoobert, Tallents, Sedwick, Martin and others have described tribes as 'aborigines' or 'aboriginals' while some others have called them as the 'primitive tribes', 'original owners' and 'jungle people' or 'forest tribes' or 'folk' respectively. In general many scholars describe tribe as a collection of families or group of families bearing a common name, members of which occupy the same territory, speak the same language and observe certain taboos regarding marriage, profession or occupation and have developed a well assessed system of reciprocity and mutuality of obligation.

In view of the overall development of some tribal groups, the Constitution of India has specified some tribes as 'Scheduled Tribes'. The specification of tribes or tribal communities as 'scheduled tribes', arose only after the constitution of India came into force. In the list of 'Scheduled Tribes', prepared in 1950, the primitiveness and backwardness were the criteria applied for specifying a scheduled tribe. The main characteristics which are common to all the scheduled tribes are:

1. Tribal origin
2. Primitive way of life
3. Habitation in remote and less easily accessible areas
4. General backwardness in all respects[4].

One third of world's tribal population live in India and ninety percent of these tribals are forest dwellers.[5]

The tribals are 'adivasis' or original dwellers, living in the subcontinent from unrecorded time and possibly driven into the forests by more aggressive invaders. Aryan groups were among the earliest to socially subjugate them. In order to escape completely and avoid domination, tribals evolved their distinct identity through endogamy, their particular practice of shifting agriculture and pastoralism, hunting and food gathering. Above all, in their intensely personal relationship with the forest around them, they formed perfectly balanced rhythms which can be described as symbiotic[6].

One third of world's tribal population live in India and ninety percent of these tribals are forest dwellers.[6] The tribal population in India has gone up from about 20 million as per 1951 census to about 66.5 million as per 1981 census. There are about 300 tribal communities living in India.[7] Numerically the most important tribal groups are Bhils, Gonds and Santals, living in central and western zone of the country.[8]

On the basis of racial features, Guha (1935) considers that the tribal population of India belongs to three races: namely the Proto-Austroloids, the Mongoloids and the Negritos.[9]

Tribal groups are homogeneous, culturally firm and have a well developed strong magico-religious health care system and they wish to survive and live in their own style.[10]

Tribal population of India has been found to speak 105 different languages and 225 subsidiary languages indicating a great deal of variety.[11] Languages spoken by Indian tribes can be classified into four major families of languages namely: Austro-Asiatic family, Tibeto-Chinese family, Dravidian family and Indo-European family.[12]

In the post-independence India these tribals along with the Scheduled castes have become the biggest minorities. They have become the base of the caste pyramid which is devoid of any staircase as currently described by Dr. Ambedhkar.[13]

Tribals in Tamil Nadu

There are 33 tribal groups living in Tamil Nadu.[14] The tribal population in Tamil Nadu is rather small in size (0.82 per cent of the total tribal population of India live in Tamil Nadu). They are also scattered over a wide area. Of the tribals in Tamil Nadu, the Palliyans and the Pulayans live in the Palni Hills.

According to 1971 census, the total population of Palliyans was 3,108 of which 3,101 (99.77 per cent) lived in the rural areas and 7 (0.23 per cent) lived in the urban areas and of the Pulayans was 4,308 of which 4,285 (99.47 per cent) lived in the rural areas and 23(0.53 per cent) in the urban areas. (The 1981, 1991 and 2001 census data apparently do not have these details).

Tribals and Forests

Traditionally, forest has been the very life supporting system of the tribals. All aspects of their economic, social, cultural and religious life are closely linked to the forest in such a way that it became part and parcel of their life. The forest dwellers depended on the forest completely for their food requirements, economic needs and medicines in times of disease. Their social life was intertwined with the agricultural cycle and forest produces. Their religion gave expression to the importance of trees, plants, animals and fruits that were essential for their socio-economic life. If the forest dwellers depended on forests for survival, it is equally clear that forests in their turn depended for their protection on the close links the forest dwellers maintained with their surroundings.[15]

To millions of tribal women and forest dwellers, forest provides fuel, fodder, food, fertilizer, medicine and raw materials such as oil, resins, gums, rubber, wax, fibre and bamboo.[16] Studies in Orissa, Madhya Pradesh, Himachal Pradesh, and Bihar indicate that up to 80 per cent of the forest families obtain between 25 per cent and 50 per cent of their food from the forest.[17]

Forests support the livelihoods of the forest dwellers in three ways:[18]

1. By providing recurrent subsistence needs—for fodder, fuel, food, fibre and many non-timber forest products
2. As a source of income
3. As capital stocks or savings bank to be cut and cashed to meet contingencies.

India's tremendous need for food and the expansion of plantations into the forest have been working together with other factors to destroy the tribals' habitat on the one hand and to offer alternative source of income on the other. They have been forced to be day labourers outside the forest.

Deprivation and exploitation by unscrupulous traders, landlords, contractors and moneylenders and people from the plains have displaced the tribals. They have lost their traditional life in the jungles as commercial industrialization ravaged the existing virgin forests.[19] Loss of forests has increased the pace of migration of tribals to the towns where they have become low paid wage labourers.[20] Deforestation has led to the destabilisation of their community living in about 5000 forest villages.[21]

Of the 14489.5 million hectares of total land area of the world, 4068.54 million ha. (30.39 per cent) are covered with forests.[22] This forest cover is disappearing at the rate of about 0.6 per cent annually. Tropical forests, by far the richest biological units in terms of genetic diversity and by far the most threatened through human activities, cover 2970 million ha. of the earth's surface (c. 20 per cent of the earth's surface housing 70 per cent of the planet's species).[23] Millions of years old tropical forests are disappearing at the rate of 24 ha. per minute,[24] closed forest at the rate of 7.8 million ha. per year and open forests at the rate of 7.8 million ha. per year.[25]

The major portion of the existing forest is owned by the Government.[26] India is losing forest cover at the rate of 1.3 million ha. every year.[27] Some 4.5 million ha. of forest land was officially diverted for agriculture and for development projects between 1950 and 1980.[28]

Besides the plant (45,000 species–12 per cent of global plant wealth) and animal (68,300 species) wealth,[29] the lives of the tribals also are greatly endangered due to deforestation as it removes the resources on which their livelihood is based.[30]

Deforestation affects these tribals more than anybody else in the following ways:[31]

1. Destruction of their traditional link with the forest and the consequent cultural alienation

2. Indebtedness, land alienation and bondage
3. Land – alienation–total lack of bargaining power, reduction in income and the lands taken over by moneylenders and middlemen
4. Bonded labour
5. Displacement and exploitation: displacement due to developmental projects such as industries, irrigation dams and mining operations
6. Physical deterioration–fall in the nutritional content of the food and consequent physical ailments and deterioration
7. The conflict between the people for whom forest is a life supporting system and forest officials for whom forest is only a raw material.

In short, the destruction of forests has resulted in impoverishment, indebtedness and in many cases land alienation and even bondage of the tribals.

The tribal folk's knowledge about the ethnobotanical wealth is still in the oral tradition. This knowledge is fast vanishing as a result of rampant deforestation and the displacement of the tribals from their traditional habitats. Hence, it is imperative for ethnobotanists to explore and conserve this folk wisdom. In this context, this research documents the ethnobotanical wealth of the tribals in the Palni hills of the Western Ghats.

REFERENCES

1. Heywood, V.H. 1992. Conservation of germplasm of wild species. In: Sandlund, O.T., Hindar, K. and Brown, A.H.D. (eds.). Conservation of Biodiversity for Sustainable Development. Scandinavian University Press, Oslo, pp. 189-203.
2. Gary and Martin, J. 1995. Ethnobotany–A methods manual. Chapman and Hall, London. pp. 268.
3. Rama Rao, N. and Henry, A.N. 1997. The Ethnobctany of Eastern Ghats in Andhra Pradesh, India. Botanical Survey of India. p. 259.
4. Chandra, Bimal. 1968. Hand Book on Scheduled Castes and Scheduled Tribes. Office of the Commissioner for Scheduled

Castes and Scheduled Tribes–Govt. of India. New Delhi. Pp. 27-28.

5. Ibid.

6. Poffenberger and Mark. 1990. Joint Management of Forest Lands: Experiences from South Asia. A Ford Foundation Programme Statement, pp. 49.

7. Ibid.

8. Maheswari, J.K. 1990. Interaction of Tribals with the Forests. Proceeding of MAB Regional Training Workshop. Tropical Forest Ecosystem Conservation and Development. South and Southeast Asia. (ed.) Nair, K.K.M., Bhat, K.V. and Sharma, J.K. Kerala Forests Research Institution, Peechi, pp. 115-120.

9. Guha, B.S. 1935. The Racial Affinities of People of India. Census of India 1931. Vol. I. Part III B. Govt. Press. Shimla.

10. Salil Basu. 1994. The State of the Art-Tribal Health in India. In: Salil Basu (ed.), Tribal Health in India. Manak Publications Pvt. Ltd. p. 312.

11. Indra Gandhi National Open University. 1990. Society in India. School of Social Sciences. ESO-02. Black 6.

12. Vidyarthi, L.P. 1983. Tribes of India. In: People of India. Some Genetical Aspect. XV. International Congress of Genetics. Dec. 12-21. ICMR. New Delhi.

13. Patil and Sharad. 1989. Mobilising Scheduled Castes and Scheduled Tribes. Economic Political Weekly, September, 2-9, 1989.

14. Government of Tamil Nadu. 1976. Sub. Plan for Tribal Development in Tamil Nadu. Government of Tamil Nadu, October 1976, pp. 181.

15. Fernandes, W. and Menon, G. 1988. Deforestation, Forest Dweller Economy and Women. Indian Social Institute, New Delhi.

16. Sharma, K. 1991. The Role of Women: The Crucial Connection. Survey of the Environment 1991. The Hindu, Madras, pp. 36.

17. Poffenberger and Mark. 1990. Joint Management of Forest Lands: Experiences From South Asia. A Ford Foundation Programme Statement, pp. 49.

18. Chambers, R., Saxena, N.C. and Shah, T. 1989. To the Hands of the Poor, Water and Trees. Oxford and IBH Publishing Co. Pvt. Ltd., New Delhi, pp. 273.

19. Akhileshwari, R. 1990. At the end of the Tether. Sunday Herald, Deccan Herald, December 30, 1990.

20. D.N. 1988. Factors in the Jharkhand Movements. Economic Political Weekly, Jan. 30, 1988.

21. NCHSE. 1987. Documentation of forests and rights. Vol. 1. National Centre for Human Settlement and Environment, New Delhi, p. 30 (mimeo).

22. Purushotham, A. 1986. Forestry in India for the 21st Century. Indian Journal of Forestry. Vol. 1. 9, No.3, pp. 183-203.

Bhatia, H.M. 1990. Planning and Forestry Development. The Indian Forests. Vol. 116. No. 81, Nov. 1990.

23. Rao, V.R. 1991. Remote Sensing: Assessment from Above. Survey of the Environment 1991. The Hindu, Madras, pp. 134-135.

24. Our Common Future. 1981. The World Commission on Environment and Development. Oxford University Press, pp. 400.

25. UNEP. 1990. The Disappearing Forest. The Environmental Sabbath. United Nations Environment Programme.

26. Lal, J.P. 1990. Deforestation: Cause and Control. The Indian Forester. Vol. 116. No. 6. June 1990, pp. 431-441.

27. Ramakrishnan, P.S. 1991. Ecology and Sustainable Development: Striking a Balance. The Survey of Environment 1991. The Hindu, Madras, pp. 148-151.

28. Sharma, K. 1991. The Role of Women: The Crucial Connection. Survey of the Environment 1991. The Hindu, Madras, pp. 36.

29. Khoshoo, T.N. 1991. Biological Diversity. A case for conservation. Survey of the Environment. 1992. The Hindu, Madras, pp. 125-127.

Gadgil, M. 1991. Conserving Biodiversity: Saving Subcontinent's Wealth. Survey of the Environment 1991. The Hindu, Madras, pp. 140-141.

30. Dasgupta and Supriya. 1988. Tribal Women. In: Understanding the Tribal Dilemma–Tribal Women and Forest Dweller Economy. Indian Social Institute, New Delhi.

31. Agarwal, Anil and Sunita, N. (eds.) 1985. The State of India's Environment 1984-85: The Second Citizen's Report. Centre for Science and Environment, New Delhi, pp. 393.

Chapter 2
Ethnobotany and its Scope

Human beings have always been largely dependent for their food, shelter, medicine and other needs on plant resources. Earlier, they were directly dependent on plants. But with the advancement of science and technology, this dependence on plants as a direct source has been slightly reduced. All the same, they remain fully dependent on plants for survival. Knowledge of this intimate relationship is the beginning of ethnobotanic inquiry.

Definition of Ethnobotany

The term ethnoecology is increasingly used to encompass all studies that describe local people's interaction with the natural environment, including sub-disciplines such as ethnobiology, ethnobotany, ethnoentomology and ethnozoology. Ethnobotany is that part of ethnoecology which concerns plants.[1]

Ethnobotany may be defined as an anthropocentric approach to botany and is essentially concerned with gathering information on plants and their uses. Power (1873-1874)[2] used the term "Aboriginal botany" to refer to a "study of all the forms of vegetable world which the aborigines used for medicine, textile, fabrics, ornaments etc". The term ETHNOBOTANY was first coined by Harshberger (1896).[3] He defined it as "the study of plants used by the primitive and aboriginal people". A truly broad concept emerged only when Robbins *et al.* (1916)[4] defined the term as a "study and evaluation of the knowledge of all phases of plant life amongst primitive societies and the effect of the vegetal environment upon the life". Vestal and Schultes (1939)[5] have employed the term ECONOMIC BOTANY "to include all plant products of use to man". They state that economic botany is more inclusive than ethnobotany.

Walter Hough (1898) defined it as "study of plants in their relation to human culture", including psychological importance and mythological reference. Barrows in 1900 inspired an investigation beyond the economic realm into religious significance of plants, their place in folk worship, folklore, etc. He was the first recipient of a doctorate in ethnobotany from the University of Chicago. Harrington (1932) considered the concept of plants in human languages, while his colleague Robinson endeavored to emphasize the influence of plants in environment in the thought, life and well-being of people.

Jones (1941)[6] expanded the utilitarian scope of ethnobotany proposed by Harshberger and others with his definition: study of "inter-relations of primitive man and plants", emphasizing that the ethnobotanists should be concerned with the entire range of relations between humans and plants as they were used in different societies for food, clothing, shelter, implements, utensils, medicines, etc. He considered ethnobotany as the unit of an ecological study specializing in their interaction in human and plant world. He changed our understanding of ethnobotany by recognizing the reciprocal and dynamic aspect of the human interaction with plants. Without a holistic and symbolic study, one cannot understand why certain plants and not others are used for a particular need, how they are prepared and when they are used, etc.

Later, Schultes (1941)[7] reiterated this point of view, stating that the men of science of one or another school are agreeing rather unanimously to regard ethnobotany as "the study of the relationship between man and his ambient vegetation". Castetter (1944)[8] confined ethnobotany to man in primitive state of culture and stated that "ethnobotany is sharply differentiated from economic botany in that it is vitally concerned with the fundamental cultural aspects of plant utilization, while economic botany practically ignores the cultural except in a very general way".

Carter (1950) envisaged ethnobotany as an ecological science capable of forging a link between geography, botany and ecology. In this respect, ethnobotany distinguishes itself from economic botany. When the former has a holistic approach to plants in relation to culture and human interaction with plants, the latter concerns more with utility potential of plants. A more modern definition of ethnobotany is given by Alcorn: "study of direct interaction between humans and plants, concerned with the totality of place of plants in a culture".

This definition gives a wider scope to ethnobotany studies. Most studies have so far not taken this broad view of the subject. Among many others, contributions of Barrau (1961), Schultes (1962), Jain (1963, 64, 65, 81, 89, 91-a,b) and Altschul (1973) are notable in understanding the scope of ethnobotany.[9]

One needs only to consult any textbook or reference book on economic botany to realize that its major interest is treating the importance of plants and the nature of plant products in relation to modern man. Therefore, ethnobotany is to be looked upon as a field of investigation which exists in its own right rather than as a branch of economic botany. Faulks (1958)[10] states that "the multitudinous connections, direct or indirect, between man and plants are proper fields of ethnobotany". In recent years, several workers defined the term in different ways. However, it can be conveniently defined as the study of the past and present inter-relationships which exist between the tribal societies and their surrounding vegetal environment.

Thus we see that over the past century, scientists of various disciplines like anthropology, archeology, economic botany, geography, pharmacology, psychology, sociology and so on have been studying the different aspects of ethnobotany. And, though a co-ordinated effort was lacking, ethnobotany did remain the central theme of several of these studies.

Ethnobotany in India

In India, this term was used by Kirtikar and Basu (1933),[11] who stated that "the ancient Hindus (Indians) should be given the credit for cultivating what is now called ethnobotany". In Vedic period the curative properties of some herbs were documented in RIGVEDA, written between 4500 and 1600 BC.[12] It was followed by ATHARVAVEDA, wherein the magico-religious aspects and utility of medicinal herbs were recorded. Later, 8 divisions of AYURVEDA, written probably between 2500 and 900 BC were the real foundation of ancient medicine. Detailed accounts of medicinal plants are also found in CHIKITSA STHANAM of SUSHRUTA SAMHITA, written between 600 BC and 300 AD.[13]

In recent years, several Indian workers like Janaki Ammal (1954, 1956 & 1975); Jain (1963, 1964, 1965, 1966, 1967, 1971, 1973, 1976, 1980, 1981, 1983 & 1991); Pal (1971, 1980, 1981 & 1986); Shah (1971, 1981 & 1982); Chaudhuri Rai (1975, 1976, 1977 & 1980); Gadgil (1976 & 1981); Tarafder (1978 & 1983); Maheswari (1981 & 1983) and Manilal (1981 & 1984) have made good contributions to ethnobotany of India.[14]

"The Indian subcontinent represents, without doubt, one of the greatest emporia of ethnobotanical wealth", says Schultes.[15] Many have ventured into this emporium to discover its wealth. Among them the first names that come to mind are E.K. Janaki Ammal and S.K. Jain. Janaki Ammal initiated the ethnobotanic study, subsistence and food plants, particularly of the tribals, as an official programme of the Botanical Survey of India. So she is considered as the pioneer in ethnobotany in India, while Jain, with his numerous books and papers on the subject, is the leading researcher in ethnobotany in India today.

The ethnobotanical studies in India were given a boost with the publication of Glimpses of Indian Ethnobotany.[16] This was the first book that gave a comprehensive view of current ethnobotanical studies in India. Publication of Watt's (1896)[17] Dictionary of the Economic Products of India was a landmark in modern ethnobotany. Wealth of India: A Dictionary of Indian Raw Material and Industrial products (1948-76), published by the Council of Scientific and Industrial Research (CSIR), has also become a reference source of ethnobotany data.

Bodding's (1925-27)[18] work on Santal Disease and Santal Medicine was a very good ethnobotanic work. Glossary of Indian Medicinal Plants by Chopra *et al.* (1956), Bibliography of Ethnobotany by Jain *et al.* (1984), Dictionary of Indian Folk Medicine and Ethnobotany by Jain (1991b), Medicinal Plants of India by Jain and De Filipps (1991) are some of the more important ethnobotanical publications of recent times. Besides these, there are several other ethnobotanists in India who have published extensively on ethnobotany of various regions in India.

The Society of Ethnobotanists (SEB), with its headquarters in Lucknow brings out a journal 'Ethnobotany'. This is the only international journal on ethnobotany till date. This SEB also conducts well designed training programmes in ethnobotany. The institute of Ethnobiology, started at Lucknow in 1995 under the Directorship of Dr. S.K. Jain is a giant leap in ethnobiological research.

Though much has been done in ethnobotanic research in India, there is a lot of scope for it, particularly on integrated and interdisciplinary lines. There is a need for a closer interaction between ethnobotanists and scientists of ethnic groups who have preserved much useful ethnobotanic information.

Among the several approaches to ethnobotanical research, the most important ones are through archaeological literature, herbarium and field surveys. The field approach plays a vital role because of the direct contact that can be established with the aboriginals/tribals living in harmony with surrounding vegetation which can easily bring out the authentic information on the uses of plants both wild and cultivated.

There are four major interrelated endeavours in ethnobotany.

1. Basic documentation of traditional botanical knowledge
2. Quantitative evaluation of the use and management of botanical resources
3. Experimental assessment of the benefits derived from plants, both for subsistence and commercial ends
4. Applied projects that seek to maximize the value that local people attain from their ecological knowledge and resources.[19]

Ethnobotany as an Emerging Science

Until recently "ethnobotany" meant only a list of plants and their uses found in the tribal, rural and other environments inhabited by aboriginal or underdeveloped people, but most of these investigators did not concentrate on the 'context' in which the 'uses' of these plants became relevant. They did not recognize that the human perception of their environment necessarily involves a complex whole when compared to the perception of the animals of their surroundings. Human encounter with nature, particularly plants, is guided besides the survival instinct, also by various values or priorities that they inherit from the society in which they are born and in which they live. Inclusion of this ethnocentric aspect into the study of botany has greatly widened the scope of ethnobotany. One important component of ethnobotany is the study of medicinal plants.

Medicinal Plants

Since the beginning of civilization, people have used plants as medicine. Perhaps as early as Neanderthal man, plants were believed to have healing powers. In fact, ancient Egyptians believed medicinal plants to have utility even in the after life of their pharaohs. Not just among primitive people but in advanced civilizations such as those of the Greeks and Romans, and even in recent times, plants have been thought to have supernatural attributes that make them effective medicines.

When pain or injury or disease struck, early man had little choice but to turn to plants. Developed empirically, by trial and error, many herbal treatments were nevertheless remarkably effective. The new medical science is reaffirming much of the old herbal lore and extending the horizons of botanical medicine.

Even now, in an age dominated by scientific and technological marvels, by miracle drugs and miracle cures, botanicals or their synthetically derived equivalents account for the majority of prescription and nonprescription medicines. Today about 25 percent of all prescription drugs are derived from higher (flowering) plants and some 12 per cent more come from lower plant forms (the molds that produce penicillin and other antibiotics are the best known examples).

Medicinal Plants and India

It has been estimated that out of about 2000 drugs that have been used in curing human ailments in India, only about 200 are of animal origin and a similar number are of mineral origin. The rest *i.e.* about 1500 are of plant origin.

Nearly 50 per cent of India's 17,000 flowering plants are medicinally used in the traditional systems of medicine and in tribal lore, where information is passed on from one generation to the other orally. Again, about three quarters of the plant drugs traded in the world today occur in natural state in India.[20]

The estimated world trade in medicinal plants gathered from our forests as Non-Timber Forest Products (NTFPs) is US $ 16 billion per year. The centre for monitoring the Indian Economy estimated that India exported plant based drugs and pharmaceutical products worth Rs.28000 millions during 1996-97 and as experience shows, they are bound to increase further.

The Indian subcontinent, one of the most distinct biogeographic regions of the world, is having a rich repository of medicinal plant resources. According to a reasonable estimate, around 70 per cent of Indian population still rely on herbal medicines for various ailments. Apart from classical systems practiced *viz.* Ayurveda, Siddha and Unani, innumerable local folk medicinal traditions are also existing. The World Health Organisation (WHO) has listed over 21,000 plant species that have been reported for their medical uses world over. In India, about 2,500 plant species belonging to more than 1,000 genera are used by traditional healers and about 500 plant-species are utilized by 159 different pharmaceutical companies.[21]

Out of an estimated 17,000 higher plant species occurring in India, about 2500 species are used over several centuries in the traditional systems of medicine *viz.* Ayurveda, Siddha, Unani and Amchi.[22] The villages and tribal folks spread across the length and breadth of the country make use of around 7500 plant species through oral traditions.[23] Thus, in addition to allopathy, the non-allopathic systems of medicines continue to be an important sector in the delivery of primary health care.[24] Tribal communities opine that ethnomedicine is more effective, harmless and appropriate for the body than modern medicine.[25]

India with her 45000 plant species and 550 tribal communities belonging to 160 linguistic groups inhabiting in varied geographic and climatic zones with diversified plant species, varied culture, rich traditional knowledge system and wisdom possess an ethnobotanical emporia. Living close to nature, the tribal communities are the custodians of unique traditional knowledge system and wisdom about ambient flora and fauna and rich heritage of Phytomedicine and Ethnomedicine.[26]

Out of 3200 taxa so far known as medicinal plants in India, a total of 343 species under 298 general and 98 families are known to be used as medicines by four major tribal groups namely Santal, Munda, Oraon and Lodha. Out of 98 families, 78 belong to dicotyledons, 18 to monocotyledons and the rest are pteridophytes and fungi. Most of these plants are utilized in preparation of herbal drugs used in traditional systems of medicine such as Ayurveda, Unani, Sidha and Tantra therapy.[26]

This study of the medicinal systems of tribal societies needs urgent attention, considering the influence of various agents which are inducing rapid changes in this section of the population. Most of the tribal societies are no longer in complete isolation from other human groups and rapid social changes can be noticed due to the contact with the peoples of other cultures. Now a number of tribals are following the life-style of the non tribal societies and their traditional norms, values, social institution and subsistence pattern have undergone vital transformation. So steps must be taken to study the ethnomedical practices of the tribals and to record all the efficacious medicines used by them, before they are lost. These should be documented before they are forgotten.

References

1. Gary and Martin, J. 1995. Ethnobotany–A methods manual. Chapman and Hall, London. pp. 268.
2. Powers, S. 1873-1874. The place of ethnobotany in ethnopharmimetic drugs. Calif. Acad. Sci. Proc. 5: 373-379.
3. Harshberger, J.W. 1896. The Purpose of Ethnobotany. Bot Gaz. 21: 146-158.
4. Robbins, W.W., Marrington, J.P. and Freire–Marreco, B. 1916. Ethnobotany of the Tewa Indians, Smithsonian Institution. Bur. Amer. Ethnol. Bill. 55.
5. Vestal, A. and Schultes, R.E. 1939. The economic botany of Kiwa Indians as it relates to the history of the tribes. U.S.A.
6. Jones, V.H. 1941. The nature and scope of ethnobotany. Chronica Botanica. 6(10): 219-221.

7. Schultes, R.E. 1941. La ethnobotanica: Su abcane y sus objectos. Caldasia 3: 7-12.

8. Castetter, E.F. 1944. The domain of ethnobiology. Amer. Nat. 78: 158-170.

9. Barrau, J. 1966. Guidelines for a study of medicinal plants used in folk medicine in Tropical Oceanic–Preliminary report to the South Pacific Commission.

Barrau, J. 1957. Edible yams of the South Sea Islands, species present, vernacular names and distribution. Kew Bulletin. 4: 309.

Schultes, R.E. 1962. The Role of Ethnobotanist in the search for new medicinal plants. Lioydia, 25(4), pp. 257-266.

Jain S.K. 1963 b. Magico-religious beliefs about the plants among the tribals of Bastar district, Madhya Pradesh. Quart. J. Myth. Soc. 54: 73-94.

Jain S.K. 1981. Ethnobotonical Research Unfolds New Vistas of Traditional Medicine. In: Glimpses of Indian Ethnobotany, (ed.) Jain S.K. Oxford & IBH Publishing Co., New Delhi, pp. 13-36.

Jain, S.K. 1964 b. Wild plant food of the tribals of Bastar (Madhya Pradesh). Proc. Nat. Inst. Sci. India 30 B(2): 56-80.

Jain, S.K. 1963. Bull. Regional Research Lab., 1, 126.

Jain, S.K. 1963a. Studies in Indian Ethnobotany–Plants used in Medicine by the tribals of Madhya Pradesh. Bull. RRL. 1: 126. 129.

Jain, S.K. 1963c. Observations on ethnobotany of the tribals of Madhya Pradesh. Vanyajati 11(4): 177-183.

Jain, S.K. 1964a. An indigenous water-bottle. Indian Forester 90: 109.

Jain, S.K. 1964c. The role of a botanist in folk-lore research. Folklore 5: 145-150.

Jain, S.K. 1965a. Medicinal plant-lore of the tribals of the Bastar. Econ. Bot. 19: 236-250.

Jain, S.K. 1965b. Studies in Indian ethnobotany-less known uses of fifty common plants from tribal areas of Madhya Pradesh. Bull. Bot. Surv. India. 5: 223-226.

Jain, S.K. 1965c. Wooden musical instruments of the Gonds of central India. Ethnomusiol. 9: 39-42.

Jain, S.K. 1967. Ethnobotany: Its scope and study. Indian Mus. Bull. 2(1): 39-43.

Jain, S.K. 1967. Plants of Indian medicine and folk-lore associated with healing of bones. Indian. J. Orthopaedics 1: 95-104.

Jain, S.K. 1968. Medicinal Plants. National Book Trust, India. pp. 211.

Jain, S.K. 1971. Some medico–religious beliefs about plants among Adibasis of Orissa. Adibasi 12: 39-44.

Jain, S.K. 1981a (ed.) Glimpses of Indian Ethnobotany. Oxford & IBH Publishing Co., New Delhi.

Jain, S.K. 1981b. Ethnobotanical research unfolds new vistas of traditional medicine. In: Jain S.K. (ed.). Glimpses of Indian Ethnobotany 13-36. Oxford & IBH Publishing Co., New Delhi.

Jain, S.K. 1991. Dictionary of Indian Folk Medicine and Ethonobotany. Deep Publications. New Delhi.

10. Faulks, P.J. 1958. An Introduction to Ethnobotany. Monedale Publications Ltd. London.

11. Kirthikar, K.R. and Basu, B.D. 1933. Indian Medicinal Plants. 4. Vols. Lalit Mohan Basu, Allahabad.

12. Bhandari, C. 1951. Vanaushadhi Chandradaya (in Hindi). Varanasi.

Pal, D.C. 1981. Plants used in treatment of cattle and birds among tribals of Eastern India. In: Jain, S.K. (ed.). Glimpses of Indian Ethnobotany 245-257. Oxford & IBH Publishing Co., New Delhi.

13. Bishagratna, K.K. 1907-1916. An English translation of Sushruta Samhita. Calcutta.

14. Janaki Ammal, E.K. 1954. The scope and functions of the reorganised Botanical Survey of India. Sci and Cult. 20: 275-280.

Janaki Ammal, E.K.1975. While the men went hunting, Indian Farming (Nov. issue) 48.

Janaki Ammal. 1956. Introduction to subsistence of economy of India. In: William, L. Thomas Jr. (ed.) Man's role in changing the face of the earth. 324-325. The University of Chicago Press, U.S.A.

Pal, D.C. 1980. Observations on folklore about plants used in veterinary medicine in Bengal & Orissa. Bull. Bot. Surv. India. 22: 96-99.

Pal, D.C. 1981. Plants used in treatment of cattle and birds among tribals of Eastern India. In: Jain, S.K. (ed.). Glimpses of Indian Ethnobotany 245-257. Oxford & IBH Publishing Co., New Delhi.

Pal, D.C. and Banerjee, D.K. 1971. Some less known plant foods among the tribals of Andhra Pradesh and Orissa states. Bull. Bot. Surv. India. 13: 221-223.

Pal, D.C. and Jain, S.K. 1998. Tribal Medicine. Naya Prokash, 206. Bidhan Sarani, Calcutta. pp. 316.

Pal, D.C. and Sareen, A.M. 1986. Some plants used by tribals in India for poisoning/stupefying fishes. J. Econ. Taxon. Bot. 8: 13-16.

Pal. 1981. Plants used in treatment of cattle and birds among tribals of Eastern India. In: Jain, S.K. (ed.) Glimpses of Indian Ethnobotany, 245-247. Oxford and IBH Publishing co., New Delhi.

Shah, N.C. 1981. The role of ethnobotany in relation to medicinal plants in India. In: Jain, S.K. (ed.). Glimpses of Indian Ethnobotany 69-80. Oxford & IBH Publishing Co., New Delhi.

Shah, N.C. 1982. Herbal folk Medicines in Northern India. J. Ethnopharmacol. 6: 293-301.

Chaudhuri Rai, A., Guha, E., Roy Chowdhury and Pal, D.C.,1980. Ethnobotanical uses of Herbaria–II. J. Econ. Taxon. Bot. 1: 163-168.

Chaudhuri Rai, H.N. and Pal, D.C. 1975. Ethnobotany of 'Vallatak'. J.Sci. Club 29: 106-110.

Chaudhuri Rai, H.N., Banerjee, D.K. and Guha, A. 1977. Ethnobotanical uses of Herbaria-I. Bull. Bot. Sur. India. 19: 256-261.

Chaudhuri Rai, H.N., Pal, D.C. and Tarafder, C.R. 1975. Less known uses of some plants from the tribal areas of Orissa. Bull. Bot. Surv. India 17: 132-136.

Chaudhuri Rai and Tribedi, G.N. 1976. On the occurrence of some medicinal plants in 24-Paraganas, West Bengal. Bull. Bot. Surv. India. 18: 161-165.

Chaudhuri, S.N., Dan, S.N., Pal, D.C., Tarafder, G.R. and Banerjee, K.K. 1977. Edible jungle leaves. Folklore 18(5): 161-163.

Choudhury, N.R. and Kumar, R. 1976. Demographic Profiles of the Bhils. Eastern Anthropologist V-29: pp. 273-280.

Gadgil, M. and Vartak, V.D. 1976a. The sacred groves of Western Ghats in India. Econ. Bot. 30: 152-160.

Gadgil, M. and Vartak, V.D. 1981. Sacred groves of Maharashtra–An inventory. In: Jain, S.K. (ed.). Glimpses of Indian Ethnobotany. 297-294. Oxford & IBH Publishing Co., New Delhi.

Tarafder, C.R. 1978. Some interesting new uses of the well known plant Vitex negundo. L. (Nasinda). Bull. Bot. Surv. India. 20: 176-177.

Tarafder, C.R. 1983 a. Ethnogynaecology in relation to plants part-I. Plants used for antifertility and conception. J. Econ. Taxon. Bot. 4: 483-489.

Tarafder, C.R. 1983 b. Ethnogynaecology in relation to plants Part II- plants used for abortion. *J. Econ. Taxon. Bot.* 4: 507-516.

Tarafder, C.R. 1983c. Traditional medicinal plants used by tribals of Ranchi and Hazaribagh districts, Bihar–Plants used in stomach troubles. *J. Econ. Taxon. Bot.* 4: 891-896.

Maheswari, J.K. 1983. Development of Ethnobotany–editorial *J. Econ. Taxon. Bot.* 4: i-v.

Maheswari, J.K. 1990. Interaction of Tribals with the Forests. Proceeding of MAB Regional Training Workshop. Tropical Forest Ecosystem Conservation and Development. South and Southeast Asia. (ed.) Nair, K.K.M., Bhat, K.V. and Sharma, J.K. Kerala Forests Research Institution, Peechi, pp. 115-120.

Maheswari, J.K., Singh, K.K. and Saha, S. 1981. The Ethnobotany of Tharus of Khari district, Uttar Pradesh. NBRI. Lucknow.

Manilal, K.S. 1981a. Ethnobotany of rices of Malabar. In: Jain, S.K. (ed.). Glimpses of Indian Ethnobotany, 297-307. Oxford & IBH Publishing Co., New Delhi.

Manilal, K.S. 1981b. An Ethnobotanic connection between mushrooms and dolmens. Ibid. 321-325.

Manilal, K.S. 1984. Hortus Malabaricus and the ethnobotanical knowledge of ancient Malabar. Ancient Sci. Life. 4: 96-99.

15. Jain S.K. 1981. Ethnobotonical Research Unfolds New Vistas of Traditional Medicine. In: Glimpses of Indian Ethnobotany, (ed.) Jain S.K. Oxford & IBH Publishing Co., New Delhi, pp. 13-36.

16. Jain, S.K. 1981a (ed.) Glimpses of Indian Ethnobotany. Oxford & IBH Publishing Co., New Delhi.

17. Watt, C. 1889-1896. A dictionary of the economic products of India. Vol. 6. Calcutta.

18. Bedding, P.O. 1925, 1927. Studies in Sandal medicines and connected folklore. Part I and II. Morn. Asist. Soc. Bengal 10: 1-427.

19. Lewis, W.H. and Elvin-Lewis, M.P. 1994. Basic, quantitative and experimental research phases of future ethnobotany with reference to the medicinal plants of South America. In: Ethnobotany and the Search for New Drugs, CIBA Foundation Symposium 185, Wiley, Chichester, pp. 60-76.

20. Manickam, V.S. 2002. Preface to New Millennium Seminar on Medicinal Plants–Proceedings. Center for Biodiversity and Biotechnology, Research Department of Botany, St. Xavier's College, Tirunelveli. pp. 246.

21. Chandel K.P.S., Shukla G and Neelam Sharma 1996. Biodiversity in Medicinal and Aromatic Plants in India: Conservation and Utilization. NBPGR, New Delhi. pp.1-239.

22. Natesh, S. 1999. 'Conservation of medicinal and aromatic plants in India: an overview. Proceedings of the symposium, 'State of the art strategies and technologies for conservation of medicinal and aromatic plants' held under the auspices of G-15 gene banks for medicinal and aromatic plants. Kaula Lumpur, Malaysia, September 29-30, 1997. Published by Forest Research Institute, Malaysia.

23. Pushpangadan, P., Nair, K.N. and Santhose, V. 1997. Biodiversity, an overview. In: The Natural Resources of Kerala, (eds.) Thambi, K.B., Nayar, N.M. and Nair, C.S., World Wide Fund for Nature–India, Thiruvananthapuram, pp. 130-139.

24. Sinha, K.P. 1994. Traditional Tribal Medicine and Modern Health Care System. In: Salil Babu (ed.). Tribal Health in India. Manak Publications Pvt. Ltd. p. 152.

25. Jose Boran, K. 1998. Tribal Ethnomedicine. Continuity and Change. A.P.H. Publishing Corporation. New Delhi. pp. 341.

26. Pal, D.C. and Jain, S.K. 1998. Tribal Medicine. Naya Prokash, 206. Bidhan Sarani, Calcutta. pp. 316.

Chapter 3

The Palni Hills in the Western Ghats

A. Location

The Hilly Regions of the Tamil Nadu

The hilly areas of Tamil Nadu (8° 5 and 13° 35 N & 76° 15 and 80° 20 E), one of the southern most states of India have a unique system of elevated areas of hilly western parts in the districts of Coimbatore, Madurai, Nilgiri and Salem.[1] The Eastern Ghats from Andhra Pradesh cut across the state in the north to meet the Nilgiri hill in the south, while the Western Ghats or the Sahyadri hills extend along the whole length of the state from the Arabian Sea to a distance of 80-160 kms. The Western Ghats have a rugged land mass and rolling topography with tropical ever green, semi-ever green and deciduous forests as well as savanahs intermixed with cultivated lands and settled areas in the valleys. The average height of the mountains from Kanyakumari in the south to Nilgris become elevated higher and higher. While it averages 1,220 m. in the southern regions, it rises to 2,554 m. at Mukurthi to 2,637 at Doddabetta in the Nilgiris. Nilgiris is the meeting place of both Eastern and Western

Ghats where the land remains elevated on an average at 2,000 m., thus forming a high plateau known as the Nilgiri plateau. A marked break in the Western Ghats, the 'Palghat gap' affords passage between Tamil Nadu and Kerala. The hill range which lies south of this gap is the Anaimalais and Cardamom hills. The Palnis are an offshoot of the Anaimalais from where the Kodaikanal group of hills are reachable. Other prominent groups of hills are the Javadhu hills, Shervarayans, Kalrayans and Pachamalais of the Kolli hills in Salem and Tiruchy districts.[2] Figure 3.1 contains the map of Tamil Nadu, with the altitudes.

Some Significant Features of the Western Ghats[3]

1. India is one of the 12 mega- diversity countries in the world.
2. India shares 2 per cent of land but 8 per cent of the global bio-diversity.
3. India is rich in floral wealth having 17,000 species of flowering plants of which 33 per cent are endemic.
4. 10 per cent of Indian flora *i.e.,* 1600 species are restricted in distribution-present only in Western Ghats.
5. The Western Ghats is one of the 18 'hotspots' of the world. (In India we have 2 'hotspots' namely Eastern Himalayas and Western Ghats).
6. Western Ghats holds about 4000 flowering plants of which 1600 species are endemic.
7. About 10 per cent of the Western Ghats flora are considered to be medicinal plants.

Topography

The Western Ghats lie between 20-40° N and 73-77° E and cover a distance of about 1600 km starting from Tapti valley in Gujarat to Kanyakumari in Tamil Nadu.These series of hills are traversing the states of Gujarat, Maharastra, Goa, Karnataka, Kerala and Tamil Nadu. There is a vast discontinuity in the Palghat gap, splitting the Nilgiri from Anamalais. The mountains rise to average altitudes between 900 and 1500 m above Sea level. The hills rise to a towering peak of 2695 m (Anaimudi) and 2637 m (Doddabetta). The Western side of the ghats, being on the threshold of the south-west monsoon (June–August) receives the maximum rainfall. The leeward side in

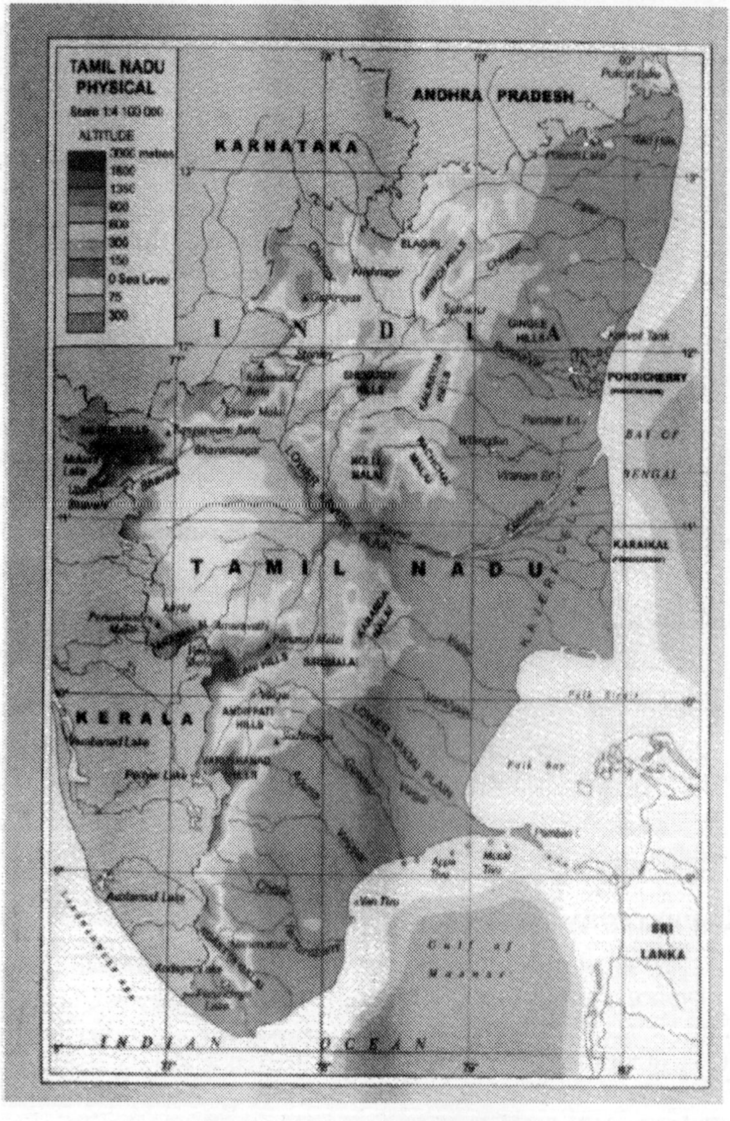

Figure 3.1: Map of Tamil Nadu

the east is the rain shadow area with much less rain. The annual rainfall varies from 2350 mm. in the north to 7450 mm. in the south. The temperature varies from 24° C in north to 20° C in south. The hottest months are April and May whereas December and January are the coldest. The varied climate and diverse topography create a wide array of habitats that support unique sets of both plant and animal species.

Floristic Wealth of the Western Ghats

The Western Ghats, South India is with high floristic diversity attracting many a botanist.[4] Wilson (1988)[5] has estimated about 30-50 million species to exist on earth, of which, only 1.7 million species have been identified so far. More than 60,000 of the 2,65,000 or so plant species are currently on the verge of extinction.[6] In India, about 45000 plant species are known to occur, of which 17000 are flowering plants and 1500 endemics.[7]

The Europeans came to the tropics in quest of spices. They were greatly fascinated by the medicinal herbs after knowing the value of the indigenous herbal medicines. The richness and diversity of tropical flora amazed the more inquisitive Europeans when they came to India. This is well evidenced from the first book in Indian botany "Coloquios dos simples" by Garcia de Orta (1563). Hortus malabaricus was published (1678-1703) by Heinrich van Rheede, which is a monumental work on the plants of Malabar. In the middle of 19th century, a series of books with illustrations on South Indian plants were published by Robert Wight. This was followed by Hooker (1872 & 1897)[8]; Gamble & Fischer (1915-1936)[9]; Cooke (1901-1908) etc. They have made significant contributions to the knowledge of the flora of the Western Ghats.

The floristic diversity may be evident from the fact that among the 17,000 species of flowering plants estimated in India, nearly 4000 are found in the Western Ghats. Sharma *et al.* (1984)[10], have enumerated 3924 spps. of flowering plants which include 1323 genera and 199 families from Karnataka. From Nilgiris alone 2611 species of flowering plants have been recorded. The Tirunelveli hills with an estimated area of about 2000 sq. kms is reported to have about 2000 spps. The diversity of the flora of Tirunelveli district is succinctly put as "an epitome of the flora of the Madras state". Nearly 150 new species and 15 genera have been reported during the last few decades from the Western Ghats and such findings tend to be an unending process. This may indicate that many more novelties are still in store for future researchers. Nearly 30 new taxa have been described from Agastyamalai region alone. Figure 3.2 presents the map of the Western Ghats.

The Palni Hills

The Palni (formerly Pulney) hills in Dindigul district are an eastward spur of the Western Ghats of India (maximum east-west length of 65 km, and north-south width of 40 km) situated between 10°12' and 10°15' N and 77°26' and 77°33' E, covering an area of

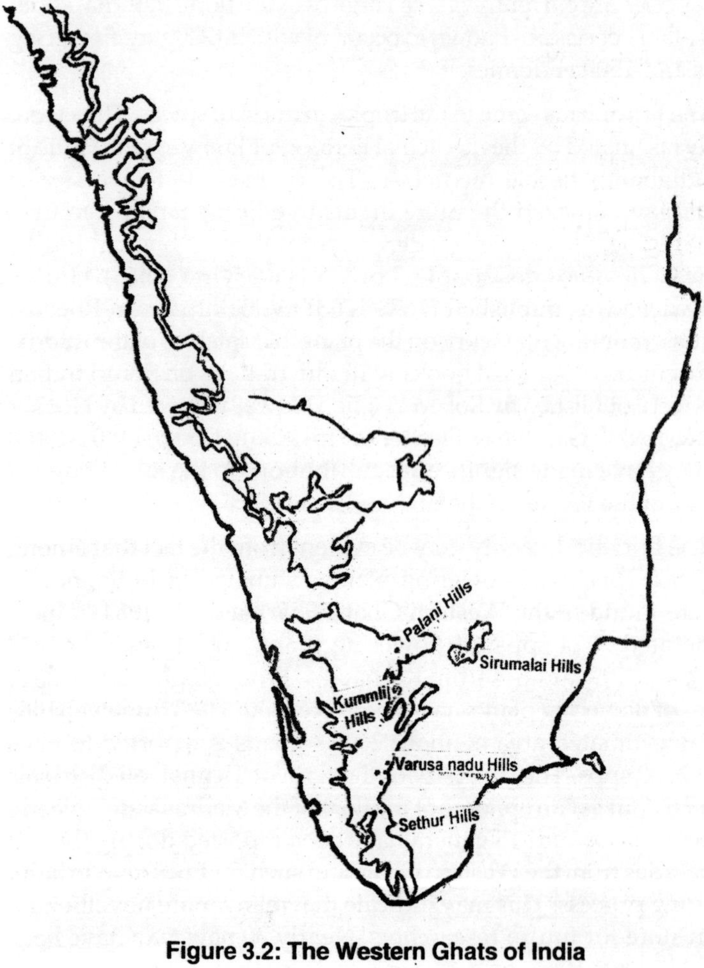

Figure 3.2: The Western Ghats of India

2068 sq.km. The uplift of the Palni hills like that of the Western Ghats and their eastern offshoots, Nilgiris and Anamalais began in the pliocene age about five million years ago.[11] The sketch of the Palni hills is given in Figure 3.3. Kodaikanal is a tourist resort in the Palni hills.

The name Kodaikanal (seven possible etymologies: The forest at the end (of the ranges), 'Forest of creepers', 'Forest green in summer', 'Gift of summer', 'Forest of umbrellas', 'Umbrella jungle', 'Fruit hills') is fortunately one among the less devastated hill stations in the country but now under severe threat of 'development'.[12]

The name Kodaikanal is applied both to the town (made a municipality in 1899 and a township in 1960) and to the taluk (constituted in 1899: area 1050 sq.km.) of which it is the headquarters. Both form part of the Palni hills.[13]

Kodaikanal is a premier hill station of southern India, sought after eagerly by the foreign and Indian elite for its healthy climate; by the people of the plains, as a refuge from nature created plagues and man-made cruelties; by the businessmen for the commercial value of its forests. Among the hill stations in India such as Simla, Darjeeling, Srinager and Ooty, only Kodaikanal has so far preserved its natural beauty[14] but this natural beauty is now being mercilessly ravaged by the commercial exploitation of the hills.

Physiography

The Palni hills are subdivided into the Upper and the Lower Palnis, two distinct geographical and even more importantly, vegetational regions, by the neutral saddle, a ravine running between the towns of Periakulam in the south and Palni in the north along the Parappar-Tevandarai Valley.

The Western block (Upper Palnis) is the plateau (c.385 sq.km. and average altitude of 2220 m) consisting of undulating slopes of grasslands interspersed with forests, locally called 'sholas' with subtropical or temperate vegetation. This is interspersed with occasional peaks rising to a maximum of 2506 m. at Vembadi but with a few ravines or valleys. The natural environment in the Upper Palnis was maintained until recent years without undue exploitation of natural resources but it is now covered with extensive commercial forestry of exotic Pine, Eucalyptus and Wattle.[15] Besides, a large area

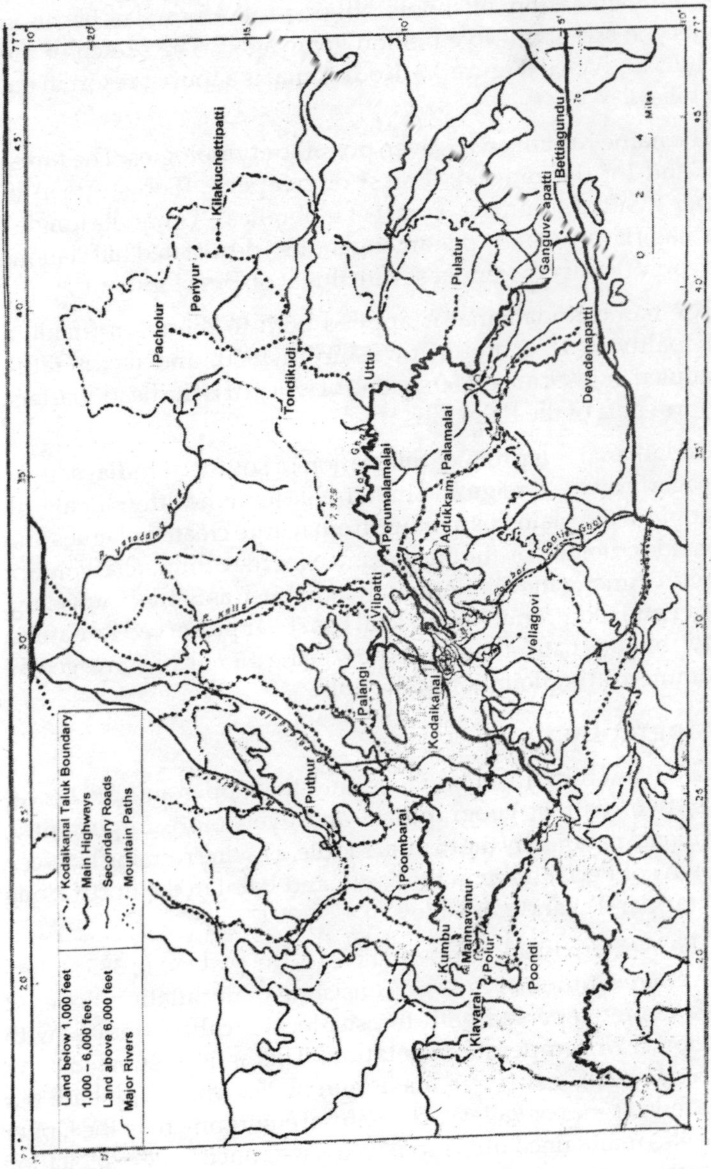

Figure 3.3: The Palni Hills

was cleared for agricultural practices, involving mainly potatoes, fruits and vegetables. The eastern block (Lower Palnis, altitude from 750-1500 m, area c. 1683 sq. km.) consists of slopes with more tropical vegetation.[16] Coffee estates are dominant in the Lower Palnis.

Kodaikanal Taluk and Township

By the turn of the century, Kodaikanal had become a recognized hill station with a considerable resident population. The first houses began to be built here in 1845. With the completion of the Ghat road of 45 km., began the inflow and influx of population.[17]

Kodaikanal taluk is under Dindigul district. It consists now of 15 village panchayats and one town panchayat, having a population of 74,619 and a township having a population of 27,461 (1991 census). There are two ghat roads connecting Kodaikanal to the plains. One is the Law's ghat road (1916), from the southern side (from Budlakundu) and another is the Palni ghat road from the northern side.

Kodaikanal township is the only township in the taluk. The total area of the township is 21.45 sq. km. (having population density of 1280 according to 1991 census). The township is located around an artificially created lake called the 'Kodai Lake' covering an area of 25 ha. with a circumference of five km.

The bottom of the basin occupied by the lake was originally a marsh, with a small stream running along it. In 1863, at the suggestion of J. Cicerson who came from Savoy, a country of lakes, Mr. Vere Leving, then collector of Madurai, banked up the streams and formed a lake.[18] Kodaikanal attracts an average of 4.73 lakh tourists annually.[19]

Panchayats and the Villages in them

Table 3.1 gives the total number of the villages in Kodaikanal taluk with specific mention of tribal villages. On the whole tribal villages constitute 33.7 per cent of the villages in Kodaikanal taluk.

Population Dynamics

Table 3.2A gives the population statistics of Kodaikanal taluk with a separate column on the population of Kodaikanal township. Going by the general trend indicated by the percentage of increase per decade, the increase has been more for Kodaikanal township than Kodaikanal taluk. Table 3.2B gives the population statistics of 2001 for Kodaikanal taluk.

Table 3.1
KODAIKANAL TALUK–PANCHAYATS AND THEIR VILLAGES

		Tribal Villages				
Sl.No.	*Panchayat Name*	*Total Villages*	*Paliyans*	*%*	*Pulaiyans*	*%*
1.	Adukkam	8	4	50.0	—	—
2.	Kamanur	9	2	22.2	—	—
3.	K.C.Patti	29	2	6.9	9	31.0
4.	Kumbaraiyur	2	—	—	—	—
5.	Kookal	4	—	—	—	—
6.	Mannavanur	5	—	—	—	—
7.	Pachalur	16	7	43.8	4	25.0
8.	Poomparai	1	—	—	—	—
9.	Poolathur	2	—	—	2	100.0
10.	Poondi	6	1	16.7	—	—
11.	Periyur	34	—	—	9	26.5
12.	Thandikudi	4	1	25.0	1	25.0
13.	Vadakavunji	6	2	33.3	—	—
14.	Vilpatti	10	2	20.0	—	—
15.	Vellakavi	3	2	66.7	—	—
16.	Pannaikadu (Town panchayat)	14	6	42.9	—	—
17.	Kodaikanal (Township)	1	—	—	—	—
		154	29	18.8	23	14.9

Sources: Panchayat Union records; Taluk office records.

Action Groups: PEAK, SIDT, KEDS and VARIED records.

Note:

1. The total number of villages includes the panchayat headquarters, town panchayat and the township.
2. All the tribal villages are predominantly populated by them except in Polur where other castes are predominant.

Table 3.2A
POPULATION STATISTICS OF KODAIKANAL TALUK

Year	*Persons*	*Total Population % Increase/ Decade*	*Kodaikanal Township*	
			Persons	*%*
1901	18,355		1,912	
1911	20,486	11.61	2,906	51.99
1921	21,582	5.35	4,283	47.38
1931	27,329	26.63	6,523	52.30
1941	33,698	23.30	9,724	49.07
1951	40,250	19.44	10,941	12.52
1961	50,710	25.99	12,860	17.54
1971	71,453	40.91	16,473	28.09
1981	83,583	16.98	20,461	34.31
1991	1,02,080	22.13	27,461	34.31

Source: The Gazetteers of India, Madurai district

The Primary Census Abstract – Kodaikanal taluk.

The Census Department, Madras.

Climate

Among the hill stations in India, Kodaikanal is said to record the lowest maximum temperature in summer and the highest maximum in winter. Never too hot and never too cold, Kodaikanal ever extends a warm welcome to all seeking a haven of refuge from the sweltering heat of the plains in summer.[21]

A curious feature of the Palni hills and even its highest reaches is that unlike most of the other areas of the Western Ghats, the region is basically dependent on the North-East monsoon and not the South-West monsoon. There are four seasons:[22]

1. Dry season–December to March (rain is scarce)
2. Warm season–April to June (frequent rains)
3. South–West monsoon season–June to September
4. North–East monsoon season–October to Novermber (about half of the annual rainfall is received during this season)

Table 3.2B
POPULATION OF KODAIKANAL TOWNSHIP AND TALUK – 2001 CENSUS

Kodaikanal Taluk	*Town*	*Total Rural/ Urban*	*Persons Males/ Females*	*Total Population*	*Total Workers Main and Marginal*	*Main Workers*	*Marginal Workers*	*Non Workers*
		T	Persons	100674	52558	47426	5132	48116
			Males	51428	31579	29319	2260	19849
			Females	49246	20979	18107	2872	28267
		R	Persons	58347	35594	32203	3391	22753
			Males	29875	19665	18248	1414	10210
			Females	28472	15929	13955	1974	12543
		U	Persons	42327	16964	15223	1741	25363
			Males	21553	11914	11071	843	9639
			Females	20774	5050	4152	898	15724
Pannaikadu	TP	U	Persons	9396	4847	4099	748	4549
			Males	4776	2771	2508	263	2005
			Females	4620	2076	1591	485	2544
Kodaikanal	M	U	Persons	32931	12117	11124	993	20814
			Males	16777	9143	8563	580	7634
			Females	16154	2974	2561	413	13180

Note: TP: Town Panchayat; M: Municipality; T: Total; R: Rural; U: Urban.

The maximum temperature in summer is 19.8° C and the minimum 11.3° C and in winter, the maximum is 17.3° C and the relative humidity is 76 per cent and the average foggy days number about 73/year. Appendix 3 gives the rainfall data in detail.

Soil

The soils of the area have been classified as broadly belonging to the class of red soils common to South India, formed by the weathering of rocks in situ. The rocks in the Palnis are a special kind of gneiss called charnockite, consisting of mica, feldspar and blue coloured quartz in varying proportion. If we look at a typical soil profile, we would find a blackish topsoil (mainly organic matter), varying in thickness from a few millimeters to about 90 cms with a porous layer underneath of yellow to red soils upto one metre in thickness which again lies above the parent rock, which however can be seen on the surface in many places. In the depressions are deeper accumulation of peaty black soil often poorly drained, and of quit low fertility.[23]

The Natural Vegetation

On the Lower Palni

Meher-Homji (1988)[24] has described in detail the vegetation zonation along the road going from Budlakundu (altitude 240 m) to the complete range of the Lower Palnis (altitude 1500 m). At the base of the hill at an altitude of 300 m. is found the thorn forest community of *Albizia amara* with *Acacia caesia, A. horrida, Cissus quadrangularis, Dichrostachys cinerea, Euphorbia antiquorum, Securinega leucopyrus, Hugonia mystax, Pterolobium hexapetalum, Sarcostemma brunonianum, Notonia grandiflora, Pterolobium hexapetalum, Sarcostemma acidum, Tarenna asiatica* and *Ziziphus xylopyrus*. A little higher up (about 500 m) appear *Gyrocarpus jacquini* and *Cohclospermum religiosum. Givotia rottleriformis* continues upto 800 m.

From 800-900 m upto 1300 m on the slopes, dominate, the species typical of the dry deciduous forest of Peninsular India. Physiognomically, they present the aspect of open forest or savanna woodland. Floristically, these are essentially composed of *Anogeissus latifolia, Albizia lebbeck, A. odoratissima, Bridelia retusa, Bischofia javancia, Buchanania lanzan, Careya arborea, Cipadessa fruticosa, Dalbergia latifolia, D. paniculata, Elaeodendron glaucum, Emblica officinalis, Garuga gamblei, Gmelina asiatica, Grewia tiliifolia, Lannea coromandelica, Pterocarpus marsupium, Santalum album, Shorea talura, Terminalia chebula, Vitex altissima, Wrightia tinctoria,* and a dwarf-palm *Phoenix humilis* var. *pedunculata.* Teak (*Tectona grandis*), a usual member of deciduous forest, is absent here but it does well in plantations. The valleys and depressions are characterised by evergreen species of wetter ecology: *Alstonia venenata, Celtis wightii, Heynea trijuga, Nephelium longana, Solanum giganteum, Trema orientalis* and deciduous *Garuga pinnata* and *Toona ciliata.*

Aditional species recorded at an elevation of 1300-1500 m are *Alseodaphne semecarpifolia, Antidesma menasu, Canarium strictum, Glycosmis cochinchinensis, Ligustrum perrottetii, Litsea deccanensis, L. sebifera, Macaranga indica, Mallotus philippensis, Meliosma microcarpa, Viburnum acuminatum* and *V. punctatum.*

On the Upper Palni

Matthew (1990)[25] has given extensive description of the flora in the Upper Palni. These areas have undulating hills originally covered with grasslands, and with dense patches of evergreen forests locally called 'sholas'.

The rolling down of the undulating plateau were originally covered with grasslands to about three-fourths of the area of the Upper Palnis. The alphabetical list of genera are given here (where more than one species is involved, such are indicated in brackets after the name).

Anagallis, Anaphalis (spp.), *Anemone, Bupleurum* (3), *Cerastium, Circaea, Crotalaria* (spp.), *Curculigo, Cyanotis, Drosera, Hedyotis* (3), *Indigofera, Knoxia, Leucas* (several), *Lysimachia* (2), *Micromeria, Murdannia, Pedicularis, Polygala, Potentilla, Ranunculus* (3), *Satyrium, Sopubia* (2), *Stellaria, Atriga, Valeriana* and *Wahlenbergia.* Marshy tracts have, besides the ubiquitous sedges, *Dipsacus, Eriocaulon* (several), *Gentiana, Hydrocotyle, Hypericum* (3), *Impātiens* (spp.), *Juncus* (4), *Laurembergia, Parnassia, Prunella, Rumex, Útricularia* (several) and *Xyris.*

The following creeper / stragglers, charaeteristic of shola border, also should be mentioned: *Clematis* (3), *Polygonum, Rosa* and *Rubus* (3). The lianous *Ficus amplocarpa,* with figs over 6 cm across, clustered on shola trees is noteworthy.

Sholas are patches of dense isolated woods composed of evergreen trees occurring along the folds on the undulating plateau, often associated with streams in the interior. These are characteristically stunted, mostly under 15 m, profusely branched and supporting a large number of epiphytes like lichen, mosses and ferns in addition to numerous creepers especially along the periphery of sholas. Among the more dominant genera (the number within brackets in the following list refers to those with more than one species) are *Syzygium* (3), *Ilex* (3), *Turpinia*, *Meliosma* (2), *Elaeocarpus* (3), *Eurya*, *Gordonia*, *Litsea* (2), *Michelia*, *Phoebe* (2), *Rhododendron* and *Daphniphyllum*. The last two, though native, are more easily noted as pioneer species along shola border when there has been disturbance of vegetation. The pride of place among shola trees goes to *Syzygium densiflorum* (*arnottianum*), known as malai naval (hill jamon). *Michelia nilagirica*, endemic to South India, locally known as Shenbagam is another noteworthy member.

The significance of the sholas, besides their unique aesthetic value lies in the following:[26]

1. They are the watershed for the surrounding plains: retaining a third of the rain water, they release the water gradually in sponge-like fashion.
2. Sholas also have considerable genetic diversity: the wild relatives of a number of cultivated plants grow here and they are indispensable for any future work on the concerned cultivated plants.

But these sholas are now being overtaken by commercial plantations and monocultures. The sholas, if once disturbed, can never stage a return to the forest physiognomy because of the inimical eco-climatic factors that prevail in the degraded open areas, having termed the existing relicts of the sholas as 'living fossil' forests. Their progressive disappearance means gradual loss of genetic resources and reduction of biological diversity.[27] Total endemics to Palni hills are thirty-nine.[28] They are greatly threatened by deforestation.

Fauna in the Palni Hills

In the account given by Krishnamani (1950)[29], the following fauna have been recorded from the Palni hills: ordinary monkey, nilgiri languor, bison, sloth bear, tigers, leopards, jungle cat, brown palm-civet cat, mongoose, sambhar, the jungle sheep ibex, wild pigs, porcupines, wild dogs, otters, shrews, hedge-hogs and malabar squirrel are some of the mammals found here. Among the birds, red-vented bulbul, the rufous-backed shrike, the Indian hoopoe, the scavenger vulture, the common kite, the laughing thrush, the fairy blue bird, the malabar trogan, the great horn-bill and the malabar grey horn-bill, the Indian and edible nest swiftlet are found here.

Exploitation of Natural Resources

The rich native vegetation of the Palnis, sheltering a variety of wildlife, was responsible for its reputation as a hill station. Their very riches led to their own destruction. Increased road communication brought in more tourists and rendered removal of hill produce easy. But by far the most destructive factor was the invasion by exotic flora, now the dominant part of the vegetation. Began to be introduced unobstrusively from the middle of the 19th century as ornamentals, fruit-bearing or commercial species by the overseas settlers on the hills, the exotic flora began to dominate gradually. Among these are the commercial Wattles, Eucalyptus, Conifers and Alder and a number of very aggressive weeds. Fruit trees and agricultural species follow closely.[30]

Land Classification

Table 3.3 gives an account of the land use classification in Kodaikanal taluk. Of the total geographical area, forests occupy c.62 percent. The Forest Department plantations are also included in this.

Plantations and Cultivated Crops

The forest department plantations include *Acacia mearnsii, Eucalyptus globulus, E. grandis, E. citriodora, Pinus patula, Tectona grandis, Alnus nepalensis, Bambusa* and *Santalum*. The cultivated fruits are *Citrus, Pyrus communis, Musa paradisiaca, Prunus salicina, Ananas comosus, Artocarpus heterophyllus, Psidium guajava, Prunus persica, Mangifera indica* and *Malus sylvestris*. The cereals grown are *Oryza sativa, Triticum vulgare, Panicum miliare, Sorghum vulgare, Eleusine coracana, Setaria italica* and *Panicum miliaceum*. The other crops cultivated are *Coffea arabica, Allium sativum, Elettaria cardamomum, Pelargonium graveolens, Camellia sinensis, Piper nigrum, Theobroma cacao, Zingiber officinale, Tamarindus indica* and *Areca catechu*. The vegetables and tubers are *Solanum tuberosum, Daucus carota* var. *sativa*, specise of *Phaseolus, Pisum, Brassica oleracea* var. *capitata, Brassica oleracea* var. *rapa* and *Beta vulgaris*. The plants cultivated for fibres are *Ceiba pentandra* var. *pentandra* and *Gossypium*.[31]

Table 3.3
LAND USE CLASSIFICATION IN KODAIKANAL TALUK

Sl.No.	*Classification*	*2000–2001 (Area in Hectares)*	*2001–2002 (Area in Hectares)*	*Difference*
1.	Forests	58919	58919	–
2.	Barren and uncultivable land	15790	15790	–
3.	Land put to non agricultural uses	10710	10708	(+) 2
4.	Cultivable waste	41	50	(-) 9
5.	Permanent pastures and other grazing land	435	437	(-) 2
6.	Land under miscellaneous tree crops and groves not included in net area sown	1130	1111	(+) 19
7.	Current fallows	103	200	(-) 97
8.	Other fallow lands	271	387	(-) 116
9.	Net area sown	16884	16681	(+) 203
	Total	104283	104283	

Source: Asst. Directorate of Statistics, Dindigul District Headquarters, Dindigul.

B. Survey Methodology

The field work commenced in December 1999 and ended in May 2002. To collect the data on the cultural aspects, the method of 'participant observation' where the observer participates in the work and livelihood of the people in the selected villages[32] was used. Living amidst them in one of their houses, sharing their food and lifestyle and sometimes participating in their work (in the estates, domestic works, NTFP collection etc.) were the main features of the stay in each one of the villages. A lot of details were observed and noted down, since the tribals, mostly illiterate, find it extremely difficult to articulate things clearly.

First 2-3 days in each tribal hamlet/village were utilized for gathering information on tribal customs, beliefs, taboos, religious rites, food habits, agricultural practices etc., from the village head and local people and these were cross-checked, critically analysed and documented.

General details were collected from all the tribal households in all the surveyed villages. An already prepared simple questionnaire, covering some areas was field tested, revised and used in the interviews. Each house in the village was also given a code. This was mainly to have some kind of order in the data and to facilitate clarity in the analysis of the data at the end.

The questionnaire was subdivided into six parts. The following is a brief summary of the structure of the questionnaire used:

1. Community particulars
2. Identification particulars and demographic details: clan, religion, deity, family members, literacy level and employment
3. House details: ownership of house, colony, patta and house structure
4. Land ownership: patta, purampokku, bhinami lands owned
5. Health: frequently affecting diseases and the different approaches for cure, utilization of the governmental health facilities
4. Cultural aspects: important festivals and ceremonies, cultural talents, food habits.

C. Data on the Plant Wealth

Various investigators like Altschul (1968)[33]; Barrau (1957[35], 1966[34]); Friedberg (1969)[36]; Jain (1964[37], 1965[38], 1967[39]); Jones (1941)[40]; Martin (1971)[41]; Scuttles (1960[42], 1962[43], 1963[44]); Turnor and Bell (1971)[45]; Vidal (1961)[46] and Von Reis (1962)[47] have suggested various tools for study of the plants used by the tribal communities. These tools were useful in the collection of data on the plant wealth.

The tribal people are usually very unwilling to disclose their knowledge about the uses of plant wealth especially except for the medicinal properties of the plants. In general, they maintain secrecy about the use of certain medicines. There is a belief that the medicines will lose their healing power if too many heads know about them. The medicinemen do not divulge a prescription in the presence of local villagers.

Resource persons for the collection of data on medicinal plants, NTFPs and other plants were carefully chosen based on the following criteria:

1. Their familiarity with plants and their uses
2. Their willingness to share information, and
3. Their ability to represent the specialist to treat certain categories of diseases.

Ethnobotanical data were collected according to the methodology suggested by Jain (1964[48], 1965a[49]) and Maheshwari (1987)[50]. Data were mainly collected through two methods:

1. Interviews in informal style with the Vaidhyar (healer), and
2. By accompanying them into the forests

Discussions were made at times with local chiefs, priests and herbal doctors not only for gathering information but for confirming the uses of same plants recorded from different informants at different places. Processing and preservation techniques of edible and medicinal plants were also recorded. The information was recorded through an audiocassette recorder on the spot.

The data were considered accurate only when 3 to 5 resource persons corroborated the information.

Authenticity of information–once the information on a particular plant was taken as reliable after repeated verification, its local name and uses were recorded. Details about the part utilized in preparation of the medicine, the ailments, the preparation, doses and prescription were also recorded.

During the field work, special care was taken to record the data on phenology, habit and habitat. The collected specimens were poisoned in a saturated solution of mercuric chloride in ratified spirit. Thereafter, they were pressed and herbarium specimens were prepared according to the conventional methods.

The voucher specimens were also collected for all the data recorded. These were identified in Rapinat Herbarium at St. Joseph's college in Trichy, using the following keys:

1. Government of Tamil Nadu. 1993. List of more common plants of Tamil Nadu and their uses. The Statistics and Planning Cell of the Office of the Chief Coservator of Forests, Madras.
2. Lushington, A.W., 1915. Vernacular list of trees, shrubs and woody climbers in Madras Presidency, Superintendent, Government Press, Vol. II-B. pp. 103.
3. Matthew, K.M., 1981-1988. The Flora of the Tamil Nadu Carnatic. Vol. I-IV. The Rapinat Herbarium, St.Joseph's College, Thiruchirapalli.
4. Gamble, J.S., 1915-1936. Flora of the Presidency of Madras. Vol. 1-3. (Vol. 3 by C.E.C. Fischer) Adlard & Sons. Ltd. London.

The voucher specimens for the plants have been given numbers and are deposited for future reference in Botany Research Lab, in St. Xavier's College, Palayamkottai in Tamil Nadu. Nomenclature of each species has been brought up-to-date as per the rules given in the International Code of Botanical Nomenclature.

The secondary data were collected from the following Government departments:

1. Adi-Dhravidar and Tribal Welfare Department, Madras–5 (mainly to get the population details of the tribals and the welfare programmes and plans)
2. District Adi-Dhravidar Welfare Office, District Headquarters, Dindigul (to get the tribal welfare programmes)
3. Department of Census, Madras (for the different census details)
4. District Land and Survey Department, Dindigul (to get the land use pattern details)
5. Forest Department, Kodaikanal (to get the complete details of NTFPs, their sale and leasing)

6. Panchayat Union Office (to find out the different welfare programmes funded and carried out by this office in Kodaikanal taluk for the tribals)
7. Agriculture and Horticulture Department in Kodaikanal (for finding out the local varieties of the cultivated crops in the tribal belts).

References

1. Nair, N.C. and Henry, A.N. 1983. Flora of Tamil Nadu Series 1: Analysis. Vol. 1. Bot. Survey of India, Howrah.
2. Matthew, K.T. 1991. Taxonomic studies on the Flora of the Palni (Pulney) Hills, Tamil Nadu II. Thesis submitted to the Bharathidasan University for the Degree of Doctor of Philosophy. The Rapinat Herbarium, St. Joseph's College (Autonomous) Tiruchirapalli, pp. 246.
3. Annamalai, R. and Ramachandran, V.S. 2002. 'Medicinal Plants Conservation Strategies in the Tropical Genepool Garden at Nadugani, Nilgiris' by Tamil Nadu Forest Department: in New millennium Seminar on Medicinal Plants Proceedings. Centre for Biodiversity and Biotechnology, India. p. 41.
4. Nair, N.C. and Daniel, P. 1986. The floristic diversity of the Western Ghats and its conservation: a review. Proc. Indian Acad. Sci.(Anim. Sci/Plant Sci.)Suppl. 127-164.
5. Wilson, E.O. 1988. The diversity of life. In: H.J. de Blij (ed.) Earth' 88 Changing Geographic Perspective, National Geographic Society, Washington, DC. pp. 68-81.
6. Roy, P.S. and Ravan, S.A. 1994. Habitat Management for Biodiversity Maintenance Using Aerospace Remote Sensing. In: Tropical Ecosystems–A synthesis of tropical ecology and conservation. (eds.). M. Balakrishnan, R. Borgstrom and S.W. Bie. Oxford–IBM Publishing Co. Pvt. Ltd. pp. 309-346.
7. Nayar, M.P. 1989. In situ conservation of wild flora resources. Bull. Bot. Surv. India 29: 319-333.

Sharma, B.D., Balakrishnan, N.P., Rao, R.R. and Hijra, P.K. 1993. Flora of India. Botanical Survey of India. Calcutta.

8. Hooker, J.D. 1872-1897. The flora of British India. Vol. 1-7. L. Reeve & Co. London.

9. Gamble, J.S. 1915-1936. Flora of the Presidency of Madras. Vol. 1-3 (Vol. 3 by C.E.C. Fischer) Adlard & Sons. Ltd. London.

10. Sharma, B.D., Balakrishnan, N.P., Rao, R.R. and Hajra, P.K. 1993. Flora of India. Botanical Survey of India. Calcutta.

11. Krishnan, M.S. 1968. Physiographic Characteristics of Peninsular Ranges. In: Mountains and Rivers of India Law B.C. (ed.). 21st Int. Geogr. Congr. New Delhi, pp. 88-95.

Vaidyanathan, R. 1977. Recent Advances in Geomorphic studies of Peninsular India. A Review of Indian Journal of Earth Science: S. Ray, pp. 13-35.

12. Matthew, K.M. 1990. The Flora of the Upper Palnis. Kodaikanal Boatclub Centenary Commemoration Vol., pp. 42-52.

13. Sambandamurthi, S. 1988. Agriculture and Eco-development in Upper Palni Hills. Paper Presented in the Seminar on 'The Conservation and Ecological Management of the Western Ghats Through Land Use Planning'. Kodaikanal 4,5 & 6, March, 1988, pp. 83-85.

14. Viraraghavan, M.S. 1988. Environmental Problems of the Palnis–Perceptions, Policies and Prescriptions. Bulletin, Madras Development Seminar Series, Vol. XVII., No. 11, pp. 550-678.

15. Development Alternatives. 1990. Sustainable Development Programme: Palni Hills, Tamil Nadu; India. A Proposal: Environment System, Development Alternatives, October 1990, pp. 29.

16. Matthew, K.M. 1991. An excursion of Central Tamil Nadu, India. Oxford IBH Publishing Co. Pvt. Ltd. New Delhi.

17. Matthew, K.M. 1969. The Exotic Flora of Kodaikanal, Palni Hills. Botanical Survey of India. 20(1), pp. 241.

18. Viraraghavan, M.S. 1988. Environmental Problems of the Palnis–Perceptions, Policies and Prescriptions. Bulletin, Madras Development Seminar Series, Vol. XVII., No. 11, pp. 550-678.

19. Balasubramanian, S.V. 1990. The Declining Kodai. Frontline May 26 to June 8, 1990, pp. 75-81.

20. Matthew, K.M. 1969. The Exotic Flora of Kodaikanal, Palni Hills. Botanical Survey of India. 20(1), pp. 241.

21. Krishnamani, V.S. 1950. Kodai Hills. Kumar Publication, Anandagiri, Kodaikanal, pp. 80.

22. Matthew, K.M. 1969. The Exotic Flora of Kodaikanal, Palni Hills. Botonical Survey of India. 20(1), pp. 241.

23. Viraraghavan, M.S. 1988. Environmental Problems of the Palnis– Perceptions, Policies and Prescriptions. Bulletin, Madras Development Seminar Series, Vol. XVII., No. 11, pp. 550-678.

24. Meher-Homji, V.M. 1988. The Palni Hills: Climate and Vegetation. Paper presented at the seminar on 'Conservation and Ecological Management of the Western Ghats Through Land Use Planning', Kodaikanal 4, 5 & 6, March 1988, pp. 39-42.

25. Matthew, K.M. 1990. The Flora of the Upper Palnis. Kodaikanal Boatclub Centenary Commemoration Vol., pp. 42-52.

26. Britto. 1986. Changing Face of the Palni Flora, Newsletter, PHCC, Winter 1986, pp. 1-14.

27. Meher-Homji, V.M. 1988. The Palni Hills: Climate and Vegetation. Paper presented at the seminar on 'Conservation and Ecological Management of the Western Ghats Through Land Use Planning', Kodaikanal 4, 5 & 6, March 1988, pp. 39-42.

28. Meher-Homji, V.M. 1978. Delineation of the Western Ghats: Phyto-geographer's View-Point. In: Glimpses of Ecology (eds.). Singh J.S. and Gopal, B. International Scientific Publishers, Jaipur, pp. 263-268.

29. Krishnamani, V.S. 1950. Kodai Hills. Kumar Publication, Anandagiri, Kodaikanal, pp. 80.

30. Palni Hills Conservation Council. 1987. Report on the Palni Hills National Park Proposal, pp. 6.

31. Matthew, K.M. 1990. The Flora of the Upper Palnis. Kodaikanal Boatclub Centenary Commemoration Vol., pp. 42-52.

32. Singleton, R., Bruce C., Margaret, M. and Ronald J. 1988. Approaches to Social Research. Oxford University Press, Oxford, New York, pp. 541.

33. Altschul, S.V.R. 1968. Useful food plants in herbarium records. Economic Botany. 22: 293-296.

34. Barrau, J. 1966. Guidelines for a study of medicinal plants used in folk medicine in Tropical Oceanic–Preliminary report to the South Pacific Commission.

35. Barrau, J. 1957. Edible yams of the South Sea Islands, species present, vernacular names and distribution. Kew Bulletin. 4: 309.

36. Friedberg, C.B. 1969. Less methods denqueteon ethnobotanique. J. Agic. Trop. Bot. Appl. 15: 297-324.

37. Jain, S.K. 1964. The role of a botanist in folk-lore research. Folklore 5: 145-150.

38. Jain, S.K. 1965b. Studies in Indian ethnobotany-less known uses of fifty common plants from tribal areas of Madhya Pradesh. Bull. Bot. Surv. India. 5: 223-226.

39. Jain, S.K. 1967. Ethnobotany: Its scope and study. Indian Mus. Bull. 2(1): 39-43.

40. Jones, V.H. 1941. The nature and scope of ethnobotany. Chronica Botanica. 6(10): 219-221.

41. Martin, M. 1971. Introduction al ethnobotany, du Cambridge, Paris.

42. Schultes, R.E. 1960. Tapping our Heritage of Ethnobotanical Lore, Economic Botany. 14: 257-263.

43. Schultes, R.E. 1962. The Role of Ethnobotanist in the search for new medicinal plants. Lloydia, 25(4), pp. 257-266.

44. Schultes, R.E.1963.The widening panorama in medicinal botany. Rhodone 65:97-120.

45. Turnor, N.C. and Bell, M.A.M. 1971. The ethnobotany of the coast Salish Indians of Vancouver Islands. Econ. Bot. 25: 63-64.

46. Vidal, J.E. 1961. Plant utilas du Lace (Monocotyledones)–V.J. Agric. Trop. Bot. Appl. (Paris). 7: 550-567.

47. Von Reis, S. 1962. Herbaria: Source of medicinal folklore. Econ. Bot. 16: 223-237.

48. Jain, S.K. 1964. The role of a botanist in folk-lore research. Folklore 5: 145-150.

49. Jain, S.K. 1965a. Medicinal plant-lore of the tribals of the Bastar. Econ. Bot. 19: 236-250.

50. Maheshwari, J.K. 1983. Development of Ethnobotany–editorial J. Econ. Taxon. Bot. 4: i-v.

Chapter 4

Sociological Aspects of Palliyans and Pulayans

The Early Inhabitants of the Palni Hills

The first people to live in the Palni Hills, who have left visible artifacts, were the dolmen-builders. Groups of dolmen circles surround Kodaikanal, which are so constructed that the entire dolmen people were thus able to communicate with each other. By 1954, Father L. Anglade of Sacred Heart College, Shembaganur, had visited and described over a hundred dolmens in the vicinity.[1]

The builders of the dolmens are unknown. Local tribes call them Pandyan forts, or hero houses, using the word hero which implies the monkey warriors who assisted Rama in crossing to Ceylon to rescue his wife Sita in the Ramayana Epic.[2] Thurston reported a theory that the present widespread agricultural Kurumba tribes are the descendants of the dolmen builders, as several groups of Kurumbas still build dolmens as resting places for their dead and others build dolmen-like temples.[3]

Previous Studies on the Palliyans and Pulayans

Within historic times atleast two primitive tribes fled into the Palni Hills from the neighbouring plains, the Palliyans and the Pulayans. The most primitive of these are the Palliyans. Early descriptions refer to them as nomadic tribes who wore grass and leaves. They lived on roots, leaves, wild dates and honey and occasionally trapped animals or fish or smoked out flying squirrels and rats from trees and holes. For cooking purposes they dug pits in the ground and lit fires over them by striking a piece of steel against a quartz stone and igniting floss from the silk-cotton tree in the spark. Some of them hired themselves out as servants to caste villagers in more recent times, where they have made a reputation of themselves by their knowledge of the medicinal properties of herbs and roots. Many of them now live permanently in villages.[4]

Marriages took place among them after puberty and they were arranged by the parents. Bride-price was much in vogue among them. Divorce can be obtained by either party on payment of a fine equal to the bride-price and widows may remarry anyone of their choice. The favourite deities of the Pulayans are Mayandi (whose shrine is generally on the knoll close to the village), Karumalayan and a goddess called Pavadai. Festivals in their honour occur in Chithirai (April-May), and consist largely in much dancing by twelve men who have sanctified themselves for the deity by abstaining from eating beef for the twelve days preceding. The whole caste is extremely fond of dancing and in Panguni (March-April) both men and women keep it up to all hours, going round and round with great energy to the sound of a drum. They eat beef, pork and even rats. They used to assemble for regular hunting excursions. The Pulayans were in the greatest subjection by their masters: the kunnuvans, who would not let them have a light at night or sleep on a cot, lent them money at exorbitant interest and turned them into slaves if they were unable to pay it back.[5]

A study was made in 1963-4 of Palliyan villages including some in the Palamalai area of the Lower Palnis, approximately twelve miles south-east of Kodaikanal.[6] The Palliyans are segmentary tribes.[7]

They have a headman with virtually no authority but a pleasing temperament, who tries to keep peace at family frictions by his words of gentle wisdom or reprimanding jokes. The major goals of the tribal units seem to be to keep peace within and avoid aggression from without. Their status among caste villagers is extremely low.[8]

The Palliyans occupy several ranges of hills in peninsular India and like the Birhors, have been classified as Australoid in physical type.[9] They are physically most similar to the Semong of Malaya and other Indian gatherers.[10] They speak Tamil but with a peculiar intonation. They are much less civilized than Pulayans. Their description as chronic 'refugees'[11] is intended to impart an appropriate understanding of their perennial retreat to avoid harassment. They sometimes build themselves grass huts, but often live on platforms up in the trees or under rocks. Their clothes are of the scantiest and dirtiest and are sometimes eked out with grass or leaves. They live upon roots (yams), leaves and honey. They cook the roots by putting them into a pit in the ground and heaping wood upon them and lighting it. The fire is kept burning all night as a protection against wild beasts and it is often the only sign of the presence of Palliyans in the jungle for they are shy folk who avoid other people. Weddings are conducted without ceremonies, the understanding being that the man shall collect food and the woman cook it.[12]

Given their character, self-sufficiency, respect for others, self-control, placation of hostile parties, and mobility are desirable; dependence, loss of self-control, unnecessary contact with strangers and provocative behaviours are to be avoided. Acts like murder, incest and theft should not be committed.[13] The nuclear family is the only unit of Palliyan society characterized by cooparation. Palliyans abstain from beef as rigidly as the most orthodox of Hindus.[14]

These studies done on the Palliyans and Pulayans mostly concentrate on the anthropological and sociological aspects of these tribals.

The Present Study

Pulayans

General Details

Pulayans, unlike Palliyans, have been living as well-knit groups in different villages ever since they occupied these hills.[15] As of now, they are living in 35 villages belonging to 11 panchayats.

On the whole there are 688 Pulayan households in the Palni Hills. Their total population, according to this present study, is 2430 with 1270 males (52.26 per cent) and 1160 females (47.74 per cent). The general trend indicated by the census data of all the villages shows that the males are more in number than the females. The biggest village in terms of total House Holds (HHs) and population is Kallakinaru which has 60 HHs with the population total of 192 (7.90 per cent of the total Pulayan population in the Palni Hills).

The major chunk of the population is concentrated between the age group 16-20 to 36-40. After age 50, there is a very drastic decline in the survivors. Very few people live beyond the age of 70.

There are a lot of deaths after the age of ten. That is why the population decreases at the age group of 11-15. Only very few females survive after the age of 65. All the above mentioned details bring out the poor health and low longevity of these people.

Clans and Deities

There are 28 clans among the Pulayans. Appendix 2A reveals that 'Budian' clan has the highest number of HHs as well as population. There are several cultural practices connected to clans. For example, a bridegroom belonging to one clan can marry a bride only from certain other clans. Specified marriage practices exist even today. Each clan has its own deity. The list of deities worshipped by the Pulayans is given in Appendix 2A. Pulayans worship 52 deities. Kudiamalaiyandi is a famous deity among them. Seventy houses keep him as their deity. Each clan celebrates the festival of its deity once a year. They also join the others in the celebrations of different local festivals.

Occupation

Most of these people are daily-wage earners. They work in the coffee estates. Of the total Pulayan population 65.23 per cent (33.74 per cent of males and 31.48 per cent of females) are daily wage earners. The occupational details are presented in Appendix 2C. The nine males who are doing Government jobs are either employed in the Indian Railways or in the government run hostels. Rampant forest destruction has displaced the tribals and has made them daily-wage earners as they can no more depend on the forests for fulfilling their food requirements. The males are paid Rs. 35 to 40 per day and the females Rs. 25 to 30 per day. The major work force is drawn from the age group of 16-45. Most of those who are unemployed are only children below the age of 15.

Education

Literacy level is very low in the case of Pulayans as in the case of Palliyans. Their villages are situated far away from schools. Even the schools closeby do not serve the interests of the tribals. Government has been running some hostels specially for tribals in the lower Palni Hills. Only 20 per cent of the tribals benefit out of these hostels. Due to economic and psychological reasons, tribal students do not continue to be these hostels. As it is seen from Appendix 2D and 2E, 70 per cent of the Pulayans remain illiterate. In this, 39.63 per cent are females who outnumber male illiterates. This is nothing unusual. It is interesting to note from the data that after fifth standard the difference in literacy level between males and females is too high. This means that most of the girls drop out after fifth standard. They go to work or stay at home.

A close look at Appendix 2D reveals that a major chunk of illiterates are from age 16-40. The drop-out rate is very high from fifth standard onwards. This shows that they cannot continue their studies as they have to earn their livelihood after the age of sixteen. Particularly at coffee seasons even ten year old girls or boys go for coffee plucking.

There are about nine girls studying 10th Standard among the Pulayans. They should be given jobs immediately after their completion. Scholarships should be given for tribals without much difficulty. Illiteracy is not only a sign of ignorance but also a vehicle of oppression and discrimination. Promotion of literacy among the tribals is an immediate need.

Housing and Land Ownership

Though the tribals have been living here for centuries, most of them do not have good shelter. A Pulayan may live in a house for years but that does not ensure his right over that house. He needs to be at the beck and call of his estate owners. More than 71 per cent of the Pulayan houses do not have pattas.

Out of 688 HHs only 660 HHs have houses. The rest live in the houses of others. This is much worse than having no ownership over the house. A person who has no house has no identity. A person who has no house, has no self image. Of the 660 HHs who have houses, only 192 HHs have pattas. These people living in these houses can be sacked at any time from there. Most of the 468 HHs live in estates. So far the Government has managed to build colony houses only for 282 HHs. Colony houses have tiled or tinned roofs. Even among those who live in colony houses, all do not have pattas.

Among the Pulayans, 260 HHs out of the total 688 HHs have no land. Even among the 260 HHs who have lands, only 132 HHs have patta lands. The rest have either Porampokku or Bhimo lands. Many would have had lands, but they have been taken away by illegal means by rich estate owners. Though the tribals are the sons and daughters of the hills, they have no house to lay their heads or no land to claim as their own. Appendix 2F gives the details of the land ownership pattern.

Palliyans

General Details

Ever since the Palliyans occupied these hills, they have been living in small groups and never in villages, whereas the Pulayans started living in villages right from the beginning. Palliyans have been kept in coffee estates as labourers. Only two decades back they started to come out of these estates and began living in bigger groups in some settlements. Now there is a growing tendency to come together and live in villages.

C. Anna Kamu who has done a detailed study on hill tribes in 1961, places Palliyan population to be around 2,000.[16] Kumbaraiyur village has both Palliyans and Pulayans. Vadagaraparai has the highest households among all the Palliyan settlements. These houses were built by Bro. James, who belonged to a non-governmental organization. These house sites were given to them in 1985 by the then sub-collector of Kodaikanal, Mr. Gurnihal Singh.

Any visitor to a Palliyan village will not find any one above the age of 70. Appendix 3A indicates that there are only a few survivors after the age of sixty. Hills are a healthy place to live. It is surprising to note that life expectancy is very low for the Palliyans. At the same time, Appendix 3A also shows a lot of people in the age group of sixteen to fifty. Women seem to outnumber men in the age group from fifteen to twenty five. Generally boys and girls begin to work from age ten onwards. By the time they reach fifty they are fully worn out. Tribals take meals only twice a day. They don't even take the minimum calories needed for a normal person. They work almost eight to ten hours a day. Tribal women, in particular, do not survive long. They work in fields and estates and look after their domestic duties. Sometimes they go to bed without food at night. So they are easily prone to all sorts of diseases. The major chunk of the population is from age ten to fifty which is the labour force. Lack of nutritious food, contaminated water, lack of medical facilities and illiteracy of the parents are the major causes of high infant mortality. Any step towards reducing infant mortality should begin with effective medical care and literacy.

Tribals are highly religious people. The only thing that they are afraid of is extra-terrestrial beings *i.e.* God and anscestral spirits. Though one may not find structured temples in tribal settlements, they are god fearing people. The Gods and Goddesses of Palliyans are mostly in forests. They are mostly forest deities (Vana Devadais). As it is specified in Appendix 3B, six out of eight deities are females. This shows that the feminine aspect of God is more appealing to them than the masculine aspect. The Goddess, 'Palichimacca', is worshiped by more people. Out of 992 people, 395 worship this deity. Next to Palichimacca, comes Goddess, Kaliamman. Kaliamman festival is one of the major festivals for Palliyans.

Occupation

The Palliyans are forced to work for more hours and paid very less. The rich estate owners exploit Palliyans and keep them as bonded labourers for generations. They have to remain in their owners' estates and work for them till they repay the loans already given.

Most of them are daily wage earners. They are paid Rs.30-35 each per day. This is insufficient for their sustenance. Next to daily wage, many people are involved in NTFP collection and oil extraction which give them better income. Unfortunately, only one person has got a government job so far. Only in Polur they are fully involved in agriculture because they own patta lands.

Education

Though government has taken pains to start schools in one or two tribal villages, they are in so many ways ineffective. So education remains even now a luxurious commodity for the tribals. As

Plate 4.1
PALLIYANS AND PULAYANS

A: Dolmen circle; B: A Palliyan family and their hut; C: A Palliyan family in Pallangi–Combai; D: A Palliyan woman of Kaduguthadi; E: A Palliyan of Kallakinaru; F: A malnourished Pulayan child

Plate 4.2

G: A Pulayan village in pristine surrounding; H: Traditional Palliyan houses; I: View of Polur–a Palliyan village; J: View of Thalingi–a Palliyan village

Appendix 3C makes it crystal clear, 91 per cent of the Palliyans are illiterate.

Though children especially girls wish to study, their economic conditions do not permit them to do so. At the age of ten itself they begin to earn or they are kept at home to look after the house works. Moreover, the existing schools do not function effectively.

As Appendix 3D indicates, there are no literates above the age of 45. Only at the age of 20-25 we see a few literates. In fact, at the age group of 6-10 we see more people going to school. This means only from last decade, Palliyans have been getting education.

House and Land Ownership

Palliyans do not waste much of their time, money and energy on building palatial houses. Rather they are happy with a roof and four walls made up of sticks and leaves. This may also be due to economic reasons. But one should not negate the simplicity with which they build their houses. Unfortunately Palliyans too do not own lands to build their houses. Most of them live in estates owned by rich people. Among the 227 households, 211 HHs have houses but only 98 HHs have pattas for their houses. Realistically speaking only 35 per cent of the Palliyans own houses. The rest are living in estates. There are about 136 families living in colony houses. There are about 135 families living in thatched houses.

Palliyans live in estates and are semi nomadic people. This is obvious from their land and house ownership pattern. Among the 277 HHs only 72 HHs have lands. The rest are landless. More than 70 per cent of the Palliyans do not have even a single piece of land. Though these tribes have been producing wealth generating crops like coffee, orange and cardamom and so on, their life remains poor and malnourished. The land ownership details are given in Appendix 3E.

The Palliyans have the following clans: Kalluppalayan, Melmalai Paliyan, Sarugu Paliyan, Theivapaliyan, Kattu Paliyan, Elai Paliyan, Theni Paliyan, Neer Paliyan, Ammana Paliyan and Theikkula Paliyan.

Plates 4.1 and 4.2 give some general details about the Pulayans and Pallians.

References

1. Anglade, L. 1954. Dolmens and Tombs of the Pulney Hills. The Anthropological Society. Bombay.
2. Victor Rosner. 1959. 'Dolmens in the Annamalai Hills, South India. In: Anthropos: International Review of Ethnology and Linguistics, LIV. 181.
3. Edgar Thurston. 1909. Castes and Tribes of Southern India, IV. Government Press. p. 139, 153 and 169-70.
4. Ibid
5. Francis, W. 1906. Madras District Gazetteers: Madura. Vol. 1, pp. 353.

   Francis, W. 1930. Madras District Gazetteers: Madura. Vol. 2, pp. 156.

   Francis, W. 1933. Madras District Gazetteers: Madura. Vol. 3, pp. 71.
6. Baliga, B.S. 1960. Madras District Gazetteers: Madurai. Government of Madras, pp. 477.
7 Peter M. Gardner 1965. "Ecology and Social structure in Refugee Populations: The Palliyans of South India"–Unpublished Ph.D. Thesis, Pennsylvania State University.
8. Marshall and Sahlins, D. 1968. 'Segmentary Tribes and Chiefdoms'. In: Tribesmen, Foundations of Modern Anthropology Series. Prentice Hall, Englewood Cliffs, N.J. pp. 20-27.
9. Peter M. Gardner 1965. "Ecology and Social structure in Refugee Populations: The Palliyans of South India"–Unpublished Ph.D. Thesis, Pennsylvania State University.
10. Gardner, P.M. 1972. The Palliyans. In: Hanters and Gatherers Today (ed.) Gupta, G.R. Carlina University Press.
11. Evans, I.H.N. 1937. The Negrilos of Malaya. Cambridge University Press.

   Furer-Haimendorf, C. 1945. The Reddis of the Bison Hills: A Study in Acculturation. London: Macmillan.

   Schebesta, P. 1927. Among the Forest Dwarfs of Malaya. London, Hutchinson.

Sklat, W. and Blagden, C.O. 1906. Pagan Races of the Malay Peninsula. London, Macmillan.

12. Gardner, P.M. 1966. Symmetric respect and a memorate knowledge: the structure and ecology of individualistic culture. S. West. J. Anthrop. 22, pp. 389-415.
13. Baliga, B.S. 1960. Madras District Gazetteers: Madurai. Government of Madras, 1960, pp. 477.
14. Gardner, P.M. 1972. The Paliyans. In: Hanters and Gatherers Today (ed.) Gupta, G.R. Carlina University Press.
15. Dahmen, F. 1908. The Palliyans, A Hill-Tribe of the Palni Hills. Anthropus 3: 1, 1908, pp. 19-31.
16. Anna Kamu, C. 1961. Melmazhai Makkal (People of the Upper Hills). Sarvodhya Publications, Tanjore, pp. 186.

Chapter 5

Cultural Aspects of the Palliyans and Pulayans

Physical Features

One can see a lot of similarity between Palliyans and Pulayans in their physical features. They are generally short and dark. They have short legs with broad soles. They have round faces with huge lips, small eyes, broad nose and curly hairs.[1] Both the groups have short necks with broad and flat foreheads. They speak swiftly, lengthening the last syllable.

Recreational Hobbies

Collecting food materials like-honey, yam and greens, firewood, fodder, etc., are some of the hobbies of both the groups. They are very fond of music. At night they play musical instruments and dance. Now they have started watching films in Panchayat T.V., listening to radio and tape recorder. Youth are interested in hunting or visiting the neighbouring towns.

Both tribes enjoy hunting and playing instruments. They playwith drum and flute type of instrument. They simply loose themselves in singing and dancing. Pulayans can play 32 ragas out of this flute. Palliyans play these instruments at night on the rock, keeping fire in front of them to warm them up.

What distinguishes tribals from the rest is their close association with the forest and their knowledge of the forest. They can smell animals from far. This is specific to tribals. Palliyans and Pulayans have these characteristics.

Dress Habits

Till last decade they had been primitive in their dressing and food habits. For example till last decade the tribal women wore only sarees. They did not wear blouses. They had a particular way of covering the chest while leaving the back open. The men did not wear shirts; still back, men used only plantain tree's outer pad to cover their back and front. There are even stories of tribes wearing dresses made up of huge leaves. While the Palliyans did not have the habit of decorating themselves with hand made ornaments, the Pulayans used many of the products from trees to decorate themselves. They had necklesses made up of wooden balls. On the middle of the nostril they wore a ring. They used a particular plant as ear ring. They even had metal rings for their hands and legs. For bathing and washing of clothes they used the fruit/bark of a particular tree.

This again reveals their close dependence on forests for their day-to-day living. They have been living very close to nature. Though the tribals have been affected by the growing consumeristic world, they still depend much on the forest for their daily needs.

Food Habits

Both Palliyans and Pulayans eat cereals like rice, wheat, ragi and bamboo rice. Both are good in making curry out of unripened papaya, jack fruit, beans and carrot. Palliyans eat a lot of greens. There are many varieties of greens that they collect from the forest to eat. Since their food habits include a lot of vitamin A, people with eye defects are minimum even at old age. That is why it is hard to see blind people among them. They also eat many types of yams. They love to eat the nuts of jack fruit, either fried or boiled. Both eat all sorts of meat. While hunting, they kill any animal they find and eat it. Palliyans abstain from eating beef as rigidly as most orthodox

Hindus.[2] They also eat many types of mushrooms. Apart from all these they eat a lot of fruits, honey and so on.

Cooking and Food Habits

The only source of cooking is firewood. It is collected either from the forest or from the estates, depending on the place and consent of the estate owners. Cooking is done twice a day (morning and evening) in all the villages except in Polur where people eat thrice a day. The items of food consumed generally include rice, greens and other products they bring from the estates where they go for work.

Advance and Loan Systems

The tribals, for managing the day-to-day expenses in the household, get an advance of money every week either from the contractors to whom they sell the NTFPs collected or from the estate owners where they go for work regularly. There is apparently no interest charged for this advance but it is very much binding on the tribals in the sense that the NTFPs collected cannot be sold to any other contractor or no increase in the price can be demanded. In the same way, for daily wage earners, they cannot go to other estates for work or demand higher wages. In short, this system of advance keeps them bonded to the contractors and estate owners.

The habit of borrowing from the contractors, estate owners and professional money lenders to meet the emergency expenses, is also prevalent among the tribals. The money borrowed is used for:

1. Religious festivals: village annual festival and common celebrations like Pongal and Deepawali.
2. Social functions like marriages, death etc.
3. Sudden and/or prolonged and serious illness

The interest for such loans is exorbitant and ranges from 150 per cent to 250 per cent.

Folk Tales among these Tribals

A. Why do Palliyans Live Inside the forests?

There is a tale about the origin of these tribals. Mannadiyars, Aasaries, Palliyans and Pulayans (the four different groups living in the Palni Hills) were all brothers. They were coming from Karur towards these hills.

On the way, the Palliyans felt very tired and exhausted and could not climb the hills. So they entered the forests and started living there in small groups. But Pulayans managed to reach the top of the hills. Even the elderly people who were struggling to climb, were pulled by the ropes attached to a hole made in their noses. This is the reason attributed to the fact that Palliyans live in forests even now and Pulayans live in villages.

There is a story about the history of these people. As the four groups of people were climbing the hill, the Palliyans saw the climber–*Dioscorea bulbifera* and started digging the yam. Since they found that tasty, they started living there. So the other three groups left the Palliyans there itself and proceeded with their journey. Even now it is believed by the three groups that Palliyans should live only in forests and live on the yams. It is very difficult to bring them out of the forest. All their thinking will be oriented only towards the forests.

B. The Reason for the Grey Complexion of the Palliyans

Once Lord Siva and Goddess Parvathi were taking a stroll on the hills. They saw a Palliyan digging out a very big yam. His concentration and attention were fully focused on the digging.

Goddess Parvathi said to Lord Siva, "Let us give a boon to this man." Moved by the hard labour of that Palliyan, Siva offered him the sacred ash with the blessing for his prosperous life.

The Palliyan, because of his focused attention on the digging said, "Lord, put the sacred ash on my back." Lord Siva in anger made that itself a boon for the Palliyan. That is why, even now, it is believed that the Palliyans are blackish grey in colour.

C. Only Dead Cattle Belong to Pulayans

All the four groups–Mannadiyars, Aasaries, Palliyans and Pulayans, who had migrated into the hills from the plains, were living happily. The Mannadiyars became landlords who amassed a lot of property for them. The Aasari lived by making agricultural implements for the Mannadiyars. Pulayans were involved in cattle-grazing.

Once, it had been raining heavily for three months on the hills. Agriculture was destroyed. There was scarcity of food. One Pulayan, because of his terrible hunger, killed one calf and ate it. This news was reported to the elder brother- the Mannadiyar. Then it was decided to divide the property among the three groups.

All the cattle were in the cattle shed. Mannadiyar took his younger brother–Pulayan and asked him what cattle he wants for his sustenance. As he saw many cattle lying on the ground, the Pulayan said, he wanted all the cattle which were lying on the ground. That time, the Mannadiyar shrewdly made some loud noise with his mouth and most of the cattle stood up. Only weak, sick and old cattle were lying on the ground. The Pulayan became the owner of this useless cattle and the Mannadiyar became the owner of all the good cattle and fertile lands.

The duties of the Pulayans were determined from this. They should play drums and do all the menial works for the Mannadiyars during all their celebrations. It is very sad to note that this slavery continues even now.

D. The Property of the Pulayans

Once the elder brother–Mannadiyar–called on the younger brother–Pulayan and showed him one small curly plant and inflorescence of coconut tree and asked him what he wanted. The Pulayan unknowingly touched the small curly plant and that was given to him. It is believed that the hair of the Pulayans is curly because he touched upon this curly plant.

During division of property, all the forest products, dead cattle and dead fowls were given to the Pulayans.

All the aforementioned tales drive home the point that for centuries, the Palliyans and Pulayans were kept separate and alienated. All these tales have been forcefully in vogue to assert the superiority and supremacy of the Mannadiyars over others. It is a sad fact that the oppressed and exploited Palliyans and Pulayans have interiorized and personalized the messages of these tales and are not ready to act or go against the established conventions and practices.

E. Why Pulayans do not tie 'Thali' (Wedding Thread) During Wedding Ceremony

Once two groups of Pulayans went to buy different things for a marriage. One group went to buy 'thali' and another group went to buy the wedding saree and toe-ring. Those who bought 'thali,' died on their way back. It was believed their spirits entered the plant– *Alysicarpus vaginalis* (Rama kirai). This is attributed as the reason why Pulayans do not tie 'thali' during marriage but tie black beads around the neck of the bride as a symbol of marriage. It is believed that the fruit of *Alysicarpus vaginalis* contains a 'thali' type of an object which has been put by the spirits of the dead Pulayans.

It is also believed that only illiterate persons should eat this fruit. If a student eats this fruit then his education will be affected.

Plates 5.1 and 5.2 bring out some of the cultural elements of these tribals.

Celebrations, Rits and Rituals of the Tribals

Celebration of the Village Festivals

Festival celebrations are held once in a year for the village deity and clan deity respectively. These celebrations make them forget all their feuds and divisions and bring unity among them. A lot of alien elements like loud speakers, video films etc., have entered their celebrations these days.

Palliyans

To initiate the preparations for the village festival, village meeting is held first under the leadership of the village head. All the details are discussed and decided. Money is collected for the expenses depending on each one's capacity. If there is no unity among the people, the village annual festival is not celebrated.

One week before the exact celebration, a rope is prepared with hay stock and neem and mango leaves are inserted into that. Then it is tied at the entrance of the village. As there is normally no permanent structure at their temple, a thatched shed is put up. For the whole week, drums and flutes are played every night and women and men dance different steps in tune with the music played.

On the day of celebration, the village god man sits on a pounder and becomes possessed with the spirit of the village deity and that pounder with the god man is carried around the village. All the people worship the deity with coconut and banana. The god man dances a particular dance in front of the temple and a special kind of music is played. Fresh rice is boiled and offered to the deity. Fresh shoots of certain grams and cereals grown in mud pots are offered by women who come in a procession carrying these pots.

Different deities are called by playing different music and dances connected to these deities are also performed. In some places songs are also sung.

When the god man is possessed with the deity, he announces all the good and bad elements in the village. If the villagers are not clear about certain things, they ask the deity for explanation. All the people give an assurance to the deity that they will, henceforth, lead a good life and ask the deity to bless them all.

During this worship, some men offer a few drops of their own blood to the deity by cutting their skin with knife. Women also dance the deity dance. At the end, a special kind of music is played and drums beaten to send away the deity. All the people share in the common meal that day.

Pulayans

Kaliamman festival is celebrated as village annual festival in many of the Pulayan villages. This is celebrated for three days.

On the first day, a thatched shed with green leaves is put up near the river or stream. The god man is taken there. When a particular kind of music of the deity is played, the god man becomes possessed with the spirit. The sacred stone, symbol of 'amman' (godess) is decorated. Women dance 'kummi'–a special kind of tribal dance.

Once the decoration is over, the god man carries the stone on his head. On the way two goats are cut and offered to the deity. Before cutting the goat, sacred ash is applied on it and water is poured on its head. If it shakes its head, then that is taken as a sign of acceptance by the deity and the goat is cut at once. On the way many people especially women become possessed with the spirit of the deity and dance. The statue (stone) is placed in the temple and worshipped.

On the second day, people offer rice flour mixed with jaggery to the deity. That is blessed by the god man and distributed among the people. The meat of the goats, cut on the first day is cooked and shared by all the people. That night too, music is played and 'kummi' is danced.

Plate 5.1
CULTURAL ASPECTS OF THE TRIBALS
A: Men dancing 'kummi' (a traditional tribal dance) at night; B: As part of ritual healing–a Palliyan woman possessed by the spirit of the village deity; C: Possessed Palliyan woman announcing the fortunes of the village; D: Tribal traditional music filling the air; E: Traditional worship place of Palliyans with an alien invasion of a structured temple

Plate 5.2

F: Traditional tribal temple at Kuravanachi odai; G: Temple at Kaduguthadi (1999); H: The same temple at Kaduguthadi (2001) with the invasion of a structure; I: An alien temple structure at Kallakinaru

On the third day, those who have already made a vow, take bath in the river and carry burning mud pots and offer to the deity. In the evening, women come in a procession and offer mud pots in which they have grown fresh shoots of some pulses and cereals. At the end of the third day, they take the decorated deity stone and leave it ceremoniously in the river.

The following is a common practice in some Palliyan and Pulayan villages:

Neem leaves and mango leaves are tied alternately in different ropes. These ropes are tied for decoration and as a sign of feast celebration all over the village especially at the entrance of the village and in front of the temple. On the first day of the village feast, a bamboo stick is put up in front of the temple. Mango leaves, neem leaves and turmeric are tied on top of the bamboo stick.

On the last day of the feast celebration, women come in a procession and offer newly grown shoots of some grains, seeds and cereals like mustard seed, ragi, millet, maize, green peas and lentil. These new mud pots are placed in front of the temple. The women folk with the men dance the traditional folk dance by going in a circle. New rice is cooked with jaggery and offered to the deity. Finally the young shoots are left in the stream or river.

Rites of Passage

As in other communities, both Palliyans and Pulayans have some rituals, customs and ceremonies connected to child-birth, puberty, marriage and death.

Child Birth

Palliyans

The pregnant woman is not treated with any special care. She carries on with her normal work till the last day. Margosa oil (oil extracted from the seeds of *Azadirachta indica*) is given to the pregnant mother to ease the labour pain. The delivery always takes places in a separate shed or in an isolated place in the house. As it is believed that the delivered mother is impure, she is kept separate for thirty days. During the first three days, the delivered mother is given a restricted diet consisting of sweetened decoction of ginger, pepper and long pepper. This preparation is called 'kayam marunthu'. Some plants and plant preparations are used to remove the retained placenta. The umbilical cord and placenta are usually put in a pot and buried nearby.

During the thirty days of seclusion, they are not supposed to make use of any vessel used by others. The house is swept with green leaves. On the thirtieth day, the delivered mother cooks food for all the other family members and serves them. She does not cook during the preceeding days. Some neighbours and close relatives are called for the meal. As a sign of welcoming them, turmeric and castor oil are applied on their eyes. New saree and jaggery are given to the one who aided in the delivery.

Castor oil and powdered turmeric and root of *Aerva lanata* are applied on the eyelids of the new born child. New cloth is tied as cradle. Pounder is placed in that first and then the child is put in the cradle. Till this ceremony, the child is made to sleep only on the floor. Village head or some elderly persons in the village name the child. Generally the names of their forefathers are given. In some places, at the head side bed of the delivered mothers are placed a sickle and a broom to ward off evil spirits and forces.

Pulayans

As Palliyans, these people also keep the delivered mother with the child in seclusion. At the end of the seventh day, some simple ceremonies are performed and they are admitted back into the original house.

Puberty

Palliyans

Once a girl attains puberty, she informs the news to her mother first. The mother at once informs this news to the maternal uncle. The uncle brings branches of plants with sap and makes a fresh shed in an isolated place. The girl is made to stay in this new shed for a period of 30 days. During this period, she should not see or meet any male. Food is prepared by her parents or relatives and served to her. The stems of trees with sap and leaves of banana, wattle and coconut trees are used to put up sheds to keep the girls in seclusion.

Before entering the hut, the girl has to take along with her a bow called 'Arjunan vill,' (a bow) made of palm stick. In some settlements, instead of 'Arjunan vill,' the girl is asked to take a bundle of 'pirandai' (*Vitis quadrangularis*) to be used as a pillow.

On the 10th or 30th day, some ceremonies are performed. The maternal uncle's son (the prospective bridegroom) welcomes all the villagers with betal leaves and aricanut put in a plate. Sometimes, marriage is also held together with this ceremony.

The girl is taken to a well or a stream. Special kind of music and drum beat, suited to the occasion, are played. Women make a houling noise while the girl is being led to the well or stream. The grass '*Cynodon dactylon*' is cut and put on a plate. The girl is made to stand on that plate. The maternal uncle pours water first and he is followed by her brother's wives. After this bath, she wears a new saree bought by her uncle. Then the deity is worshipped with the offering of coconut, flowers and fruits. Water is filled in a silver container and three or five betal leaves are placed in that water. The girl is made to carry the vessel on her head without spilling the water. Cloth is spread on the way and she walks on the cloth. Sumptuous meal is prepared at home and served to all. The girl is made to collect all the banana leaves used for eating the served meal.

In some places, the girl is seated on a stone and seven pots of water are poured on her and the prospective grooms throw on her balls of cow dung laden with pods of 'erukku' (fruits of *Calotropis procera*) and charcoal, which brings an end to the ceremony associated with puberty.

Moustache is drawn for a girl who is a friend of this girl. Turban is tied on her head. She is given a bow and an arrow. She is made to wear a shirt and dhoti. This friend is made to stand with the girl who has attained puberty. Dances and drum beats with flute music are performed in front of them. Sometimes this girl and her friend are carried by some people and danced.

Pulayans

Once a girl attains puberty, that news is first reported to the maternal uncle and then to the other villagers. A thatched shed with green leaves is put up.

Iron rod or sickle and lemon are placed on the lap of the girl and she is sent into that shed. These things are placed on her lap in order to prevent the attack by evil spirits. She is sent in with one of her girl friends. She should not see her maternal uncle or any prospective grooms during these days of seclusion.

After thirty days, some ceremonies are performed. A small bronze container is decorated with saffron powder, turmeric and sandle wood. Ginger cluster, turmeric plant and mango leaves are tied around that vessel. The girl carries that on her head to a stream or a well. Music and drums are played. She is bathed there. Meal is prepared for all and served. Generally, sticks of *Chenopodium ambrosoides* and *Ficus tsjakela* are used to put up sheds to keep the girls who have attained puberty, by both the tribes.

Marriage

Palliyans

Palliyans generally marry cousins and marriages between uncles and nieces are also common. Marriage outside their own tribe is unusual. Elders decide about the marriage. Till some years back, those who like, started living together without any formal ceremony and celebration. "Thali" (yellow thread) tied around the neck of a woman and toe-rings distinguish the married women from others. "Thali" is generally a simple turmeric applied yellow twine. Their weddings are usually conducted without ceremonies.[3]

'Thali' is prepared by the bridegroom's family and toe-ring by the bride's family. Details about both the partners can be obtained well in advance. Marriages are normally held in bridegroom's place. An elderly person hands over the 'thali' thread to the bridegroom and he ties it around the neck of the bride. A special kind of drum beat and music are played for this occasion of marriage.

The marriage of widowed and divorced persons of either sex are permissible. A few Palliyans have started practicing polygamy in recent years. In general, girls get married soon after attaining puberty.

Pulayans

Marriages are generally held within the tribe. There are some restrictions and conventions not to marry among and between certain clans. First the family of the boy goes in search of prospective brides. Some ceremonies are held for the marriage.

A pole is cut out of a tree with sap and placed at the site of the marriage 'pandal' (shed). This pole is called the 'auspicious pole'. Some paddy, ragi and grains of *Setaria* with a coin, are tied on the pole. A big stone is placed in front of the bridegroom's house and that is the marriage place. On the day of the marriage, the maternal uncle places the bridegroom on that stone. Nine kinds of grains and cereals, turmeric and betal leaves are placed there. Maternal uncle hands over the 'thali' to the bridgroom and he ties it around the neck of the bride. Black beads arranged in a thread is tied as thali. There is a folk tale for this custom that has been already explained.

Then the married couple is taken to the forests where they worship their deity and after coming back to their village, they get the blessings of the elders of the village. Wedding dinner is prepared and served to all. Generally, stems of *Ficus glomerata* are used to put up the marriage pandal and sheds for any other auspicious occasion.

The following practice is prevalent in some of the Palliyan and Pulayan villages:

Eight days before the marriage, a stem from a tree with sap is put in the middle of the ground where marriage shed is going to be put up. Normally stems of *Ficus religiosa* or *Commiphora caudata* are used for this purpose. On top of this stem is tied a cloth bundle with nine different grams and grains like ragi, millet, maize, blackgram etc. Water is poured to this stem for the following eight days. *Euphorbia tirucalli* is also tied on the stem.

A new mud pot is decorated and soaked lime powder is applied outside the pot. Water mixed with ground turmeric is poured inside. A platform with sand and mud is prepared in the marriage 'pandal'. This mud pot is placed in front of this platform. The couple sits on the platform. 'Thali' is placed in a plate. Seven persons worship the sun god on behalf of the bride and six on behalf of the bridegroom. Then the 'thali' is handed over by the maternal uncle to the bridegroom and he ties it around the neck of the bride.

Some Positive Aspects in their marriage practices are:

1. Bride or bridegroom can stay in the village or house of each other to get a full knowledge about each other before marriage.
2. Acceptance is obtained from both the marrying partners before marriage.

3. Dowry is neither demanded nor given.
4. Any one can get divorce. Remarriage is permitted.
5. Love marriages are accepted.

Funeral Rites

Palliyans

The news of the death is first informed to the uncles and cousin–brothers. They come to this house and symbolically hand over the dead body to the brothers of the dead person. When all the expected people arrive, the dead body is given an oil bath. The body is taken in a procession to the burial ground. Before starting the procession, their local deity is worshipped with coconut and fruits. Drum beat and music for this occasion are played.

While going, water is sprinkled on the way and while coming back, cow-dung mixed water is sprinkled. Women do not go till the burial ground. When the funeral procession passes through a junction of some roads, seven lines are drawn and some coins are thrown. It is done with the belief that evil spirits will not attack anyone. If the deseased is a father, then the hair on the head of the younger son is shaved off. If the deseased is a mother, then the hair on the head of the elder son is shaved off. If there are no sons, then a little hair on the hand of the daughter is saved off. This daughter alone can go upto the burial ground.

All those who took part in the funeral procession, take bath and leave for their houses. If death happens due to some unnatural causes, then the body is cremated.

Pulayans

The news of the death is reported to all the villagers and relatives. They bring rice and money and offer to the bereaved family. After offering coconut and fruits to the deity, the body is taken to the burial ground in a procession. From death till burial, special kind of drum beat and music are played. Women can follow the procession upto a certain distance. If the dead person is a husband, then while giving oil bath to the dead person, his wife is also given bath and the 'thali' is removed from her neck.

Gingeli seeds and mustard seeds are sprinkled in front of the funeral procession, while going. Once the funeral procession leaves the house, plain water or cow dung mixed water is thrown or sprinkled in front of the house.

Dance and Music

Dance and music are part and parcel of these tribal groups. Particular types of dance and music exist for all occasions and celebrations.

Drum and flute like instrument are the fundamental instruments in their music. Wood of *Gmelina arborea* is used to make the drum.

Music and dances are performed for all the rites of passage, any special occasion, auspicious occasions, festivals and celebrations. Men, women and children dance together. These dances are performed generally at night. Keeping burning charcoal or wood in the middle, people move in a circle and dance.

The dance for the village deity is danced only by the 'Poojari' (god man). There are various folk songs in vogue among these people. These songs are sung on various occasions.

References

1. Anna Kamu, C. 1961. Melmazhai Makkal (People of the Upper Hills). Sarvodhya Publications, Tanjore, pp. 186.
2. Dahmen, F. 1908. The Palliyans, A Hill-Tribe of the Palni Hills. Anthropus 3: 1, 1908, pp. 19-31.
3. Thurston, E. and Rangachari, K. 1909. Palliyans. In: Castes and Tribes of Southern India. V. 461-472, Madras.

Chapter 6

Health Care and Medical Practices

Health Status of the Tribals

There is a general agreement that the health status of the tribal population in India is very poor. Different studies[1] have tried to establish this with the help of morbidity, mortality and health statistics. The widespread poverty, illiteracy, malnutrition, absence of safe drinking water and sanitary living conditions, poor maternal and child health service, ineffective coverage of national health and nutritional service have been traced out in several studies as possible contributing factors for dismal health conditions prevailing among these vulnerable population.

Though modern medicines have made inroads into their health status, they still depend on their traditional herbal remedies for cure.

The Palliyans and Pulayans are frequently affected by cold, fever and cough. They suffer much from stomach disorders like amoebic dysentery and so on. A lot of people have scabies and cut wounds. Tribals are occasionally affected by diseases like small pox, blue fever, jaundice and tuberculosis. These are the common diseases by which the tribals suffer generally.

As one can see from the list, many of the diseases are caused by lack of clean water. Though these tribes live in interior forests they do not have clean drinking water facility. Even the existing water sources are polluted by the plantations and estates. Moreover the parents do not take enough care to keep their children clean.

Till last decade they used medicinal plants to cure their diseases. But now people slowly have begun to go to allopathic hospitals. Though Government has established primary health centres here and there, they are not of use to the tribals. Generally they do not function effectively and even while they function, they do not have medicines to meet the needs of these illiterate tribes. One can see a lot of self made quak doctors going around tribal villages. They neither have scientific knowledge nor proper tools and medicines to treat the diseases.

Tribal Ethnomedicine

In the absence of explanation, diseases and calamities were attributed to wrath of gods, the invasion of the body by evil spirits or malevolent influence of stars and planets.[2] It is thus obvious that the art of healing in the prehistoric era was combined with superstition aided by administration of herbal remedies available in the vicinity.

Tribals are in general the followers of animism. It leads to the belief that diseases and death are caused by certain spirits or supernatural powers. Sometimes bigotry in faith of dead spirits who live with living bodies is in existence. These beliefs have a great influence on their attitude and psychology about the ailments. They have to take steps to counteract the inimical influence of the hidden powers, to pacify, propitiate or satisfy some and fight and punish others like witches.[3]

Hughes (1968)[4] has coined the term ethnomedicine, to denote the medical beliefs and practices found in primitive and folk societies, which is by and large accepted by anthropologists everywhere.

During the last few decades scientists all over the world are paying much more attention to the studies of newer branch of science, Ethnobiology, specially to tribal medicine or ethnomedicine. In India also scores of scientists and scholars are working in these new fields of knowledge. Since 1960s this study has been intensified. India with her 45000 plant species and 550 tribal communities belonging to 160 linguistic groups inhabited in varied geographic and climatic zones with diversified plant species, varied culture, rich traditional knowledge system and wisdom possess an ethnobotanical emporia. Living close to nature, the tribal communities are the custodians of unique traditional knowledge system and wisdom about ambient flora and fauna and rich heritage of Phytomedicine and Ethnomedicine.[5]

Professor R.E Scuttles of the Harvard University, U.S.A is of the opinion that India with her many living groups of people, having diversified ethnic culture, history of rituals and performance, who are more or less isolated from modern world and are closely associated with their ambient vegetation is the emporia of ethnobotanical research.[6]

On personal investigation among the Santal communities, Santal Parganas of Bihar, Bedding (1925,1927)[7] had brought out a magnum opus on Studies in Santal medicine and connected folklore (Part I, II, and III) and Kirtikar and Basu (1935)[8] in their Indian medicinal plants (vol.1-4) mentioned a few plants used by the tribal people. Second author of this book has made substantial contribution on phytomedicine of Gonds of Madhya Pradesh. It may be submitted that practically no comprehensive account of the tribal knowledge system on therapeutic uses of plant species has so far been published.

Prain's Bengal Plants (1903)[9] covered present Bangladesh and Chota Nagpur of Bihar and Haine's Botany of Bihar and Orissa (1921), was perhaps the first and only account of the flora. Delton (1872)[10] Hunter (1879)[11] and others have dealt with the habits, language, etc. of Indian tribal people. Bhowmik (1963)[12] has given a short account of ethnology of Lodhas of Midnapur district of West Bengal.

The knowledge of the economic plants in general and medicinal plant in parts of West Bengal, Orissa and Bihar states is meagre, except for the general work of Watt (1885-1895).[13] Moreover there is no written information on indigenous knowledge system about plant and plant parts used by the different ethnic groups in Bihar except the works of Bodding (1925,1927) and Bressers (1951).[14]

Out of 3200 taxa so far known as medicinal plants in India, a total of 343 species under 298 genera and 98 families are known to be used as medicine by four major tribal groups namely Santal, Munda, Oraon and Lodha. Out of 98 families, 78 are dicotyledons, 18 monocotyledons and the rest are pteridophytes and fungi. Most of these plants are utilized in preparation of herbal drugs used in traditional systems of medicine such as Ayurveda, Unani, Sidha and Tantra therapy.[15]

It is seen that many plants used by the tribals have important compounds which have been used in medicine and commerce. So modern medicine is paying increasing attention to the medicinal plants used by the forest dwelling people.[16]

The importance of traditional (indigenous health practices) tribal medicinal system and the like drew the attention of Bodding (1925)[17] Vidyarthi (1969)[18] Mittal (1979),[19] Pulu (1979),[20] Mathur (1982)[21] and others. Apart from these, there are a number of studies particularly on the curative aspect of plants/herbs which are used by the tribals (Jain, 1966).[22] Jain also points out the importance to document them and use them for adequate experimentation.

Several scholars[23] have highlighted the importance of tribal medicine in their studies and they insist that the use of herbal medicine should be documented before they are forgotten.

The study of the medical systems of tribal societies needs urgent attention, considering the influence of various agents which are inducing rapid changes in this section of the population. Most of the tribal societies are no longer enjoying complete isolation from other human groups and rapid social changes can be noticed due to contact with the peoples of other cultures. Now a number of tribals are following the life-style of the non tribal societies and their traditional norms, values, social institution, and subsistence pattern have undergone vital transformation. So steps must be taken to study the ethnomedical practices of the tribals and to record all the efficacious medicines used by them, before they are lost.

Tribal Perception of Etiology, Health and Medicine

According to Saraswati Swain (1994),[24] the tribal belief of causation of diseases can be broadly categorised into two areas and they are: A. Supernatural and B. Physical.

Supernatural causation includes soul loss, breach of taboo, sorcery, ghosts of ancestors, ancestral spirits and supernatural powers, evil eye and evil mouth. The physical causes include the effect of weather, effect of wrong food, accidents and natural calamities. The treatment procedure can be broadly divided into preventive and curative methods though such demarcation does not exist and the whole field of taking care is a continuous one. The preventive procedures include use of charms, amulets, animal sacrifice, propitiation of disease seeking spirit, worship of god and belief in protective function of rituals. The curative practices include first worship of deities and spirits. To the tribals, religion and medicine are not separate. Herbal medicines are given next.

The spirits of the dead ancestors in every culture play an important role in ensuring health, prosperity and protection to the family. But when these ancestral spirits are not properly honoured, worshipped or humoured, they invariably inflict some afflictions on the members of the family or cattle or crop. The tribal people devote a lot of their time, energy and material resources to honouring, worshipping and appeasing spirits, ghosts, deities, gods and goddesses.[25]

The various studies of Thurston (1907),[26] on the tribes of Travancore, Gnanambal (1955)[27] on Uralis tribe, Majumdar (1961),[28] Troisi (1979)[29] on Santal, Hockings (1980)[30] on Badagas of South India, Egnor (1984),[31] Vidyarthi and Rai (1977)[32] on Ho, Jaintia, Korwa, Maler, Oraon tribes, Mathur (1982, 1987)[33] on the tribes of Wayanad, Radhakrishnan (1986)[34] on Kanikkars of Kerala, Nair Viswanathan (1985, 1987),[35] Mahapatra (1986)[36] on Santal of Bihar, Sen (1986)[37] of Andamanese of Andaman islands, Guha (1986)[38] on Boro tribe of Assam, Battacharya and Sengupta (1986)[39] on Birhor tribe, Dash (1986)[40] on Paraja of Orissa, Razvi (1986)[41] on Jaunsari tribe of Uttar Pradesh, Chakraborty (1986)[42] on Kherwar, Chero, Kol, Mar, Oraon, Munda tribes of Bihar, Upadhyay (1987)[43] on Asur, Birgia, Kisan tribes of Bihar, Tiwari (1987)[44] on Raj, Shauka tribes of Uttar Pradesh, Menon and Ramesh (1988)[45] on tribes of Arunachal Pradesh, Elwin (1955)[46] on the Saora tribe of Orissa, Saunder (1954),[47] Simmons (1960)[48] on Mestizo tribe of Coastal Peru and Chile, Hughes (1968),[49] and Foster (1976)[50] affirm the various aspects of the supernatural and natural causation of diseases.

For the tribals, religion plays an important role in all spheres of their life. Even in health care, religious beliefs and practices have their own specific influence. Viswanathan Nair (1985)[51] has observed, "Most of the annual ancestral propitiation ceremonies can be considered as ventures to ensure proper health of the society". They also believe that if proper propitiation is not offered to the deities and to ancestral spirits, then they will get angry and send diseases and other calamities to the members of the community.

Socio-Cultural Dimensions of Health

There is a close relationship between medicines and religion, morality and magic.[52] Elwin (1947)[53] has given an inventory of gods associated with various diseases in the Saora pantheon associated with children's diseases, cough, cold, sore throat, boils, blindness, rheumatism, madness, diseases of pregnant women and diseases of animals. Most of these diseases can be cured by propitiating these gods directly or through shamans.

The wisdom of replacing the traditional system and introducing the so-called modern system in the tribal areas without critically examining their usefulness has been questioned by many scholars particularly in view of the fact that there does exist a traditional system of medicine.[54]

Knowledge and documentation of traditional tribal medicines are also urgently necessary. It is important for the modern system to begin with the base existing in the tribal society and build on it, rather than replace it. In order to achieve this, these should be documented before they are forgotten.[55]

Tribal Healers

There are two types of healers among the two tribals (Palliyans and Pulayans). They are:

1. Ritualists
2. Herbalists

Ritualists

The ritualists invoke God's blessings to cure diseases. They believe that sickness is caused by specific spirits–good and bad. Both men and women practice ritualism. The whole healing ceremony takes about a day to complete. The person who initiates and leads the proceedings of the ceremony is supposed to be possessed by good spirits. He/she chants 'mantras' (prayers) by gently rubbing a *Casuarina* stick on the body of the patients. During the process, the spirit–possessed healer causes some small bruises in his body and sheds a few drops of blood on the patient as an offering to the deity.

A man who sits next performs the other part of the ceremony by throwing a few paddy grains. Grains falling distantly from the place of this ceremony indicate that the spirits causing the diseases are willing to leave the body of the patient. If any two grains fall closely shaped like a 'T', it is interpreted that the spirit's way to go out from the patient's body is blocked and the process is repeated again till the block is removed.

Once the spirit is willing to leave, the ritualist who performs the ceremony alerts another person who is ready with cooked rice mixed separately with coal powder and turmeric made into balls. The ritualist runs for a short distance towards west and the person with the cooked rice balls runs after him. He throws the rice balls in all four directions from where the ritualist stands. This event forms the terminal part of the ceremony after which the patient is said to be free from the possession of the evil spirit. Usually, such treatment is followed for diseases such as severe rheumatism and paralysis. Often the ritualistic healers also prepare 'talisman' with herbal drugs and tie them to the patients. The aforementioned ritual practice is done if it is believed that the diseases are caused by spirits, good or bad.

In some tribal villages it is believed that failure to fulfill some promises and pledges made to the deities or bad behaviour of the villagers cause diseases especially prolonged illness and epidemics.

In this case, on a particular day, all the villagers gather around the village temple. The 'poojari' (priest) first becomes possessed with the spirit of the deity. The villagers ask the priest for reasons for the disease or epidemics. Once the causes are known which are generally some unfulfilled pledges or bad behaviours, the villagers make an assurance to the deity that they will fulfill the pledge and rectify their wrongs and sins. Unfulfilled pledges may include failure to celebrate the annual village festival, some vows made etc. Wrong doings may include unconventional sexual behaviours, enmity etc. People repent and ask for forgiveness during the ceremony. Coconuts and fruits are offered to the deity. The priest blesses the sacred ash and distributes to all the people. He also proposes some penances to appease the anger of the village god. The proposed penances are fulfilled within a stipulated time and 'pongal' (unboiled rice cooked with jagery) is prepared and offered to god at the end.

Herbalists

Both Palliyans and Pulayans are good herbalists. Their knowledge about medicinal plants is rather specialized and is limited to a few members in the community. They are recognized as 'vaidyars' or medicine men. They are generally respected the most and considered indispensable members within the tribal society. In general, the medicine men treat all kinds of illnesses but some of them are specialized in treating some specific illnesses. The remedies of common ailments like cuts, head ache, pain, fever, etc., are known to most members of these tribal communities. In each locality, there are several individuals, who, though not recognized as medicine men, possess such knowledge and act as reliable informants.

These tribals are usually not unwilling to disclose their knowledge about the uses of plants except for their medicinal properties. They maintain secrecy about the use of certain medicines for they believe the medicines will lose their healing power if too many people know about them. Some tribals have made a name for themselves for their knowledge of the medicinal properties of herbs.

From the account of the enthnobotanical data, it is clear that the Palliyans and Pulayans like other ancient tribals[56] possess the ability to discern the character of various plants and their beneficial properties. This knowledge is fast disappearing. This view is consistent with the findings of studies done on tribal communities of Andhra Pradesh[57], Orissa[58], Uttar Pradesh[59], Assam[60], Western Himalayas[61] and Bihar.[62]

Herbalists treat patients only by using plant resources. They diagnose diseases based on the symptoms told by the patients as well as on their personal experience in treating human ailments. Usually some of them are generalists who treat people for wide ranging diseases. However, there are also specialists who are adept in treating one or a few diseases such as rheumatism and paralysis.

Plate 6.1
TRIBAL HEALERS
A: Kanniappan - Perungadu; B: Mariammal - Pallangi - Combai; C: Ramuthai - Kaduguthadi; D: Muthusamy - Kaduguthadi; E: Veeramma - Pethuparai; F: Mariamma - Vattakkanal; G: Antony Cruz - Palamalai; H: Samadhanam and Marianandam - Palamalai

Plate 6.2
I: Indirani and Palaniammal - Sembadiyuthu; J: Parvathi - Nallurkadu; K: Murugamma - Kaduguthadi;
L: Muthupandy - Sembarankulam; M: Palanichamy - Thalingi

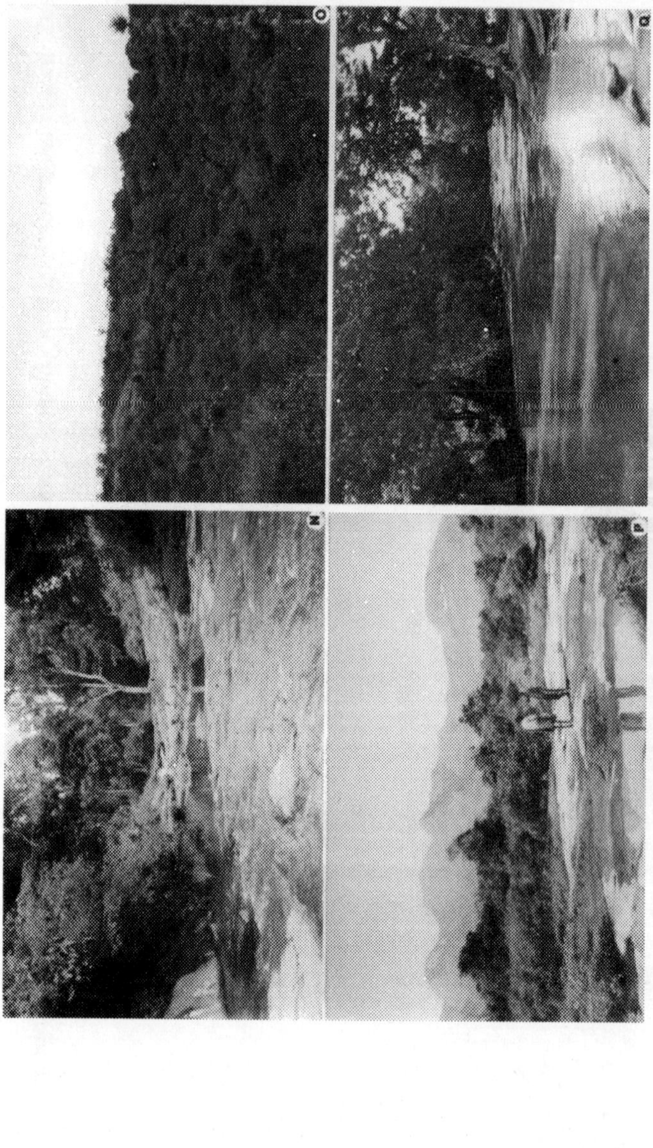

Plate 6.3

AREAS OF PLANT COLLECTION

N: Kallakinaru river; O: Dense forest in Pallangi–Combai P: Periphery of the forest in Thalingi

Q: Banks of a river with medicinal plants

The herbalists have a wide range of herbal remedies for common cold, cough, head ache, stomach problems, piles and poisonous bites and stings. Generally for cold, asthma and cough, these tribals use plant species such as *Adhatoda zeylanica, Acalypha indica*; similar report for various ethnic groups of India have been given by Shah (1984);[63] Prasad and Abraham (1984);[64] Ragupathy and Mahadevan (1991);[65] Kapur (1993);[66] Negi *et al.*, (1993)[67] and Gill *et al.*, (1993).[68]

For skin diseases, these tribals mainly use plant species such as *Acalypha indica* and *Azadirachta indica* from time immemorial. The same observation has been made among the native people of various places in India and Nigeria by Prasad and Abraham (1984);[69] Shah (1984);[70] Gill *et al.*, (1993);[71] Kapur (1993)[72] and Karuthapandi and Britto (1993).[73]

Hemidesmus indicus is generally used by these tribals for snake bite. The same plant has been used by various tribal people of Tamil Nadu, Madhya Pradesh and Andhra Pradesh.[74]

Karuthapandi and Britto (1993),[75] reported that the plants such as *Ocimum tenuiflorum, Sansevieria roxburghiana, Solanum trilobatum* and *Vitex negundo* had been used for ailments of fever, ear ache, caugh and head ache. The tribals in the Palni hills also used these plants for the same purpose.

Plates 6.1 and 6.2 present different details about tribal herbalists.

References

1. Basu and Salil (ed.) 1994. Tribal Health in India. Manak Publication Pvt. Ltd. pp. 349.

   Basu, A. 1990. Anthropological Approach to Tribal Health. In: Tribal Demography and Development in North-East India, edited by Ashish Bose, Tiplut Nagbri and Niklesh Kumar. B.R. Publishing Corporations, Delhi, India pp. 131-142.

   Basu, S.K. 1986. Genetics, Socio-cultural and Health Care among Tribal Groups of Jagdalpur and Konta Tahsils of Bestar District, M.P. In: Anthropology Development and Nation Building. (Ed.) A.K. Kalla and K.S. Singh. Concept Publishing Company, New Delhi, pp. 87-105.

   Basu, S.K., Jindal, A., Kshatriya, G.K., Singh, P., Roy, P. and Sharma, K.K.N. 1989. Epidemiological Investigation of

Haemoglobinopathies and Allied Disorders, Nutrition and Physical Growth Trends, Health Profile, Health Seeking Behaviour and Their Environmental Correlates for Promotion of Health care Among Scheduled Tribes and Scheduled Castes of Bihar District, Madhya Pradesh. Project Report. Department of Population Genetics and Hum. Dev., NIHFW, New Delhi. (DGHS).

Bardhan, A., Sinha, S.K, Gopinath, K. and Brahmbhatt, A.K. 1989. A Diagnostic study of factors responsible for less or no utilisation of health facilities among the Bhils of Madhya Pradesh–PAC Report, NIHFW, New Delhi (Mimeo).

Swain, S., Jena, S.C. and Singh, P. 1990. Morbidity Status of the Kondha Tribes of Phulbari (Orissa). In: Cultural and Environmental Dimension of Health. (eds.) Chaurdhari, B. Inter India Publication, New Delhi, pp. 177-191.

Sen, D.K. 1986. Some Notes on the Fertility of Jaunsari Women. Eastern Anthropologist V. 10, No. 1. pp. 60-67.

Mahapatra, D.K. and Das, J. 1990. Nutritional Eco-systems of Orissa Tribes. In: Buddhadeb Chaudhuri (ed.), Cultural and Environmental Dimension on Health.

Rizvi, S.N.H. 1986. 'Health Practices of the Jaunsaries–A Socio Cultural Analysis'. In: Buddhadeb Chaudhuri (ed.). Tribal Health. Inter India Publications, New Delhi, pp. 223-246.

Mukherjee, B.M. 1986. 'Aging Members and their Health in Changing Techno-Economic Condition'. In: Buddhadeb Chaudhuri (ed.). Tribal Health. Inter India Publications, New Delhi.

Haque, M. 1990. Height, Wt., and Nutrition among the Six Tribes of India. In: Cultural and Environmental Dimension on Health. (eds.) Chaudhuri, B., Inter India Publication, New Delhi, pp. 102-106.

Choudhury, N.R. and Kumar, R. 1976. Demographic Profiles of the Bhils. Eastern Anthropologist V-29: pp. 273-280.

2. Park, J.E. and Park. 1977. Textbook of Preventive and Social Medicine. Banarsidas Bhanot, Jabalpur.

3. Pal, D.C. and Jain, S.K. 1998. Tribal Medicine. Naya Prokash, 206. Bidhan Sarani, Calcutta. pp. 316.

4. Hughes and Charles C. 1968. 'Medical Care Ethnomedicine'. In: David L. Sills (ed.) International Encyclopedia of the Social Sciences, Vol. 10. The Macmillian Co. and the Free Press, New York.

5. Pal, D.C. and Jain, S.K. 1998. Tribal Medicine. Naya Prokash, 206. Bidhan Sarani, Calcutta. pp. 316.

6. Schultes, R.E. 1960. Tapping our Heritage of Ethnobotanical Lore, Economic Botany. 14: 257-263.

7. Bedding, P.O. 1925, 1927. Studies in Sandal medicines and connected folklore. Part I and II. Morn. Asist. Soc. Bengal 10: 1-427.

8. Kirtikar, K.R. and Basu, A.R. 1935. Indian Medicinal Plants (2nd ed. revised by Blatter, J.F. Caius and K.S. Maskar) Vol. 4. Allahabad.

9. Prain, D. 1903. Bengal Plants. Vol. 2. Calcutta.

10. Delton, E.T. 1872. Descriptive ethnology of Bengal, Calcutta.

11. Hunter, W.W. 1879. A statistical account of Bengal. Vols.2 (Reprinted Edition, 1973) New Delhi.

12 Bhowmik, P.K. 1963. The Lodhas of West Bengal–A Socio-economic Survey. Puthi Pustak. Calcutta.

13. Watt, C. 1889-1896. A dictionary of the economic products of India. Vol. 6. Calcutta.

14. Bedding, P.O. 1925, 1927. Studies in Sandal medicines and connected folklore. Part I and II. Morn. Asist. Soc. Bengal 10: 1-427.

15. Pal, D.C. and Jain, S.K. 1998. Tribal Medicine. Naya Prokash, 206. Bidhan Sarani, Calcutta. pp. 316.

16. Pal, D.C. and Jain, S.K. 1998. Tribal Medicine. Naya Prokash, 206. Bidhan Sarani, Calcutta. pp. 316.

17. Bodding, P.O. 1925. Santhal Medicine Memoir, Asiatic Society of Bengal, 10: 2.

18. Vidyarthi, L.P. 1969. Studies of Health and Culture in Tribal Bihar; A Preliminary Note, Paper presented at the Conference

on Review of Behavioural Research. In: Health and Extension Education, New Delhi. (Mimeographed).

19. Mittal, K. 1979. Primitive Medicines Versus Modern Medicine among Santhal. Jr. Sr. Res., 22(1): 49-54.
20. Pulu, J. 1979. The Plants: Those are used by the Idu of Lohit district for Medicinal Purpose, Risarum, 5.2: 10-14.
21. Mathur, P.R.G. 1982. Arthoropology of Tribal Medicine: Disease and Curing Techniques Among the Tribes of North Wynad (Kerala) Man in India, Vol 62. No. 3. p. 295-312.
22. Jain, S.K. 1968. Medicinal Plants. National Book Trust, India. pp. 211.
23. Mathur, P.R.G. 1982. Arthoropology of Tribal Medicine: Disease and Curing Techniques Among the Tribes of North Wynad (Kerala) Man in India, Vol 62. No. 3. p. 295-312.

    Nair, N. and Vishwanathan. 1987. 'Tribes, Healers, Medicinal Matter and Environment'. Paper presented at the National Workshop on Tribal Health and Medicine in Forest Environment. Dept of KIRTADS. Kozhikode, Kerala.

    Narayan and Ravi. 1989. "Towards a People–Oriented Alternative Health Care System". Social Action. Vol. 39. No. 3.

    Bhattacharya, S., Sengupta and Syamal Kanti. 1986. 'The Concept of Disease and its Treatment among the Birhors of Purulia'. In: Buddhadeb Chaudhuri (ed.). Tribal Health. Inter India Publications, New Delhi.

    Joseph, A.N.T. 1989. 'Ethnozoology of Reptiles in Relation to Health Care among Tribals in Madhya Pradesh'. Social Action. Vol. 39. No. 4.

24. Saraswati Swain. 1994. 'Health Disease and Health Seeking Behaviour of Tribal People of India'. In: Basu, Salil, (ed.). Tribal Health in India. Manak Publications Pvt. Ltd. p. 15.
25. Basu, A. 1990. Anthropological Approach to Tribal Health. In: Tribal Demography and Development in North-East India, edited by Ashish Bose, Tiplut Nagbri and Niklesh Kumar. B.R. Publishing Corporations, Delhi, India pp. 131-142.

26 Thurston and Edgar. 1909. Castes and Tribes of Southern India. Cosmo Publications, Delhi.

27. Gnanambal, K. 1955. 'The Religious Beliefs of the Urali'. Bulletin of the Department of Anthropology, Vol. IV. No.2.

28. Majumdar, D.N. 1933. 'Disease, Death and Dioination in Certain Primitive Societies in India. Man in India, 13: 115-149.

29. Troisi, J. 1979. Tribal Religion: Religious Beliefs and Practices among the Santals. Manohar Publications, New Delhi.

30. Hockings and Paul. 1980. Sex and Disease in a Mountain Community. Vikas, New Delhi.

31. Egnor and Margaret Trawick. 1984. 'The changed mother or what small pox goddess did when there was no more smallpox'. In: E.V. Daniel and J.F. Pugh (eds.), Contributions to Asian Studies, Vol. 1-18. E.J. Brill, Leiden, The Netherlands.

32. Vidyarthi, L.P. and Rai, B.K. 1977. Tribal Culture in India. Concept Publishing Company, Delhi.

33 Mathur, P.R.G. 1982. Arthoropology of Tribal Medicine: Disease and Curing Techniques Among the Tribes of North Wynad (Kerala) Man in India, Vol 62. No. 3. p. 295-312.

Mathur, P.R.G. 1987. 'The Ethno medicine of the Irular of Attapady, Kerala', Paper presented at the National Workshop on Anthropology of Tribal Health and Medicine in Forest Environment. Dept. of KIRTADS. Kozhikode, Kerala.

34. Radhakrishnan, V.N. 1985. 'The Traditional Medicine of Kanikkars. Kalpadhenus Vol. 13. No.3.

35. Nair, N. and Vishwanathan. 1985. 'Tribal Health and Medicine in Kerala'. Unpublished Ph.D. Thesis submitted to the Calicut University. Kerala.

Nair, N. and Vishwanathan. 1987. 'Tribes, Healers, Medicinal Matter and Environment'. Paper presented at the National Workshop on Tribal Health and Medicine in Forest Environment. Dept of KIRTADS. Kozhikode, Kerala.

36. Mahapatra and Sitakant. 1986. Modernization and Ritual: Identity and Change in Santal Society. Oxford University Press. Calcutta.

Mahapatra, D.K. and Das, J. 1990. Nutritional Eco-systems of Orissa Tribes. In: Buddhadeb Chaudhuri (ed.), Cultural and Environmental Dimension on Health.

37. Sen, D.K. 1986. Some Notes on the Fertility of Jaunsari Women. Eastern Anthropologist V. 10, No. 1. pp. 60-67.

38. Guha and Anima. 1986. 'Folk Medicine of the Borokacharis–A Plain Tribe of Assam'. In: Buddhadeb Chaudhuri (ed.) Tribal Health. Inter–India Publications. New Delhi.

39. Bhattacharya, S., Sengupta and Syamal Kanti. 1986. 'The Concept of Disease and its Treatment among the Birhors of Purulia'. In: Buddhadeb Chaudhuri (ed.). Tribal Health. Inter India Publications, New Delhi.

40. Dash and Jaganath. 1986. 'The Concept and Treatment of Disease in Tribal Orissa'. In: Buddhadeb Chaudhuri (ed.). Tribal Health. Inter India Publication, New Delhi.

41. Rizvi, S.N.H. 1986. 'Health Practices of the Jaunsaries–A Socio Cultural Analysis'. In: Buddhadeb Chaudhuri (ed.). Tribal Health. Inter India Publications, New Delhi, pp. 223-246.

42. Chakraborthy and Aniruddha. 1986. 'Some Findings on the use of Traditional Medicine in the Rural Areas of Palamaus'. In: Buddhadeb Chaudhuri (ed.) Tribal Health. Inter India Publications, New Delhi.

43. Upadhyay, V.S. 1987. 'Health in Forest Environment', Paper presented at the National Workshop on Tribal Health and Medicine in Forest Environment. Dept. of KIRTADS, Kozhikode, Kerala.

44. Tiwari, S.C. 1987. 'Health and Disease in the Tribes of Uttara Khand', Paper presented at the National Workshop on Tribal Health and Medicine in Forest Environment. Dept. of KIRTADS, Kozhikode, Kerala.

45. Menon and Ramesh. 1988. 'The Influence of Spirits'. India Today. Vol. XIII. No. 22.

46. Elwin, V. 1955. The Religion of an Indian Tribe. Oxford University Press, London.

47. Saunder and Lyle. 1954. 'Healing Ways'. In: J.D. Jennings and E.A. Hoebel (eds.), Readings in Anthropology. MC Graw Hill Book Company, New York.

48. Simmons and Ozzie G. 1960. 'Popular and Modern Medicine in Mestizo Communities'. In: Dorrian Apple (ed.). Sociological

Studies on Health and Sickness. Mc. Graw-Hill Book Company. Inc. New York.

49. Hughes and Charles C. 1968. 'Medical Care Ethnomedicine'. In: David L. Sills (ed.) International Encyclopedia of the Social Sciences, Vol. 10. The Macmillian Co. and the Free Press, New York.

50. Foster and George M. 1976. 'Disease Etiologies in Non-Western Medical Systems'. American Anthropologist -Vol. 78: No. 4.

51. Nair, N. and Vishwanathan. 1985. 'Tribal Health and Medicine in Kerala'. Unpublished Ph.D. Thesis submitted to the Calicut University. Kerala.

52. Sachidananda. 1994. 'Socio-cultural Dimensions of Tribal Health'. In: Basu, Salil (ed.). Tribal Health in India. Manak Publications Pvt. Ltd. p. 64.

53. Elwin, V. 1955. The Religion of an Indian Tribe. Oxford University Press, London.

54. Chaudhuri Buddhadeb. 1986. 'Social and Cultural Aspects of Health'. Journal of Social and Economic Studies, (N.S.). 379-388.

55. Chaudhuri Buddhadeb. 1989. Social Science, Health and Culture: The Tribal Situation, Social Action, 39(3): 243-253.

56. Rajasingh, G.J. 1971. Forest working plan for the Tirunelveli North Division, Government of Madras Publication, Madras. pp. 127-133.

57 Reddy, M.H., Vijayalakshmi, K. and Venkata Raju, R.R. 1996. Native phytotherapy for snake bite in Nallamalais Eastern Ghats, India. J. Econ. Tax. Bot. Addl. Ser. 12: 214-217.

58. Aminuddin and R.D. Girach. 1991. Ethnobotanical studies on Bondo tribe of district Koraput (Orissa). India. Ethnobotany. 3: 15-19.

59. Negi, K.S., Tiwari, J.K., Gaur, R.D. and Pant, K.C. 1993. Notes on ethnobotany of five districts of Garhwal Himalaya, Uttar Pradesh, India. Ethnobotany. 5: 73-81.

60. Borthakur, S.K. 1993. Native phytotherapy for child and women diseases from Assam. Ethnobotany. 5: 87-91.

61. Rana, T.S., Bhaskar, D. and Rao, R.R. 1996. Strategies for sustainable utilization of plant resources by the tribals of the Tons Valley, Western Himalaya. Ethnobotany. 8: 96-104.

62. Girach. R.D., and Aminuddin. 1995. Ethnomedicinal uses of plants among the tribals of Singabhum district, Bihar, India. Ethnobotany 7: 103-107.

63. Shah, G.L. 1984. Some economically important plants of Salsette island near Bombay. J. Econ. Tax. Bot. 5: 753-765.

64. Prasad, N.P. and Abraham. 1984. Ethnobotany of Nayadin of North Kerala. J. Econ. Tax. Bot. 5: 41-48.

65. Ragupathy, S. and Mahadevan, A. 1991. Ethnobotany of Kodiakkarai Reserve Forests, Tamil Nadu, South India. Ethnobotany. 3: 79-82.

66. Kapur, S.K. 1993. Ethno-medico plants of Kangra valley (Himachal Pradesh). J. Econ. Tax. Bot. 17(2): 395-408.

67. Negi, K.S., Tiwari, J.K., Gaur, R.D. and Pant, K.C. 1993. Notes on ethnobotany of five districts of Garhwal Himalaya, Uttar Pradesh, India. Ethnobotany. 5: 73-81.

68. Gill, L.S., Nyawuame, H.G.K., Esezoboar, E.I. and Osagie, I.S. 1993. Nigerian folk medicine: Practices and beliefs of Esan people. Ethnobotany. 5: 129-142.

69. Prasad, N.P. and Abraham. 1984. Ethnobotany of Nayadin of North Kerala. J. Econ. Tax. Bot. 5: 41-48.

70. Shah, G.L. 1984. Some economically important plants of Salsette island near Bombay. J. Econ. Tax. Bot. 5: 753-765.

71. Gill, L.S., Nyawuame, H.G.K., Esezoboar, E.I. and Osagie, I.S. 1993. Nigerian folk medicine: Practices and beliefs of Esan people. Ethnobotany. 5: 129-142.

72. Kapur, S.K. 1993. Ethno-medico plants of Kangra valley (Himachal Pradesh). J. Econ. Tax. Bot. 17(2): 395-408.

73. Karuthapandi, G. and John De Britto, A. 1993. Medicinal uses of plants in Cheranmahadevi hill area of Tirunelveli district in Tamil Nadu. J. Econ. Tax. Bot. 17. 361-366.

74. Jain, S.K. 1963. Studies in Indian Ethnobotany–Plants used in Medicine by the tribals of Madhya Pradesh. Bull. RRL. 1: 126. 129.

Jain, S.K. and Tarafder, C.R. 1963. Native plant remedies for snake-bite among adivasis of central India. Indian Med. J. 57: 307-390.

Jain, S.K., Banerjee, S.K. and Pal, D.C. 1973. Medicinal plants among certain Adibasis in India. Bull. Bot. Surv. India. 15: 85-91.

75. Karuthapandi, G. and John De Britto, A. 1993. Medicinal uses of plants in Cheranmahadevi hill area of Tirunelveli district in Tamil Nadu. J. Econ. Tax. Bot. 17. 361-366.

Chapter 7

Ethnobotanical Wealth of the Tribals

A total of 326 plants have been collected and identified as used by these tribals for their various needs. They are collected and used for the following purposes: medicine, food, domestic uses, culture and sale. Among the plants, *Gmelina arborea, Carissa carandas, Phyllanthus emblica* and *Mangifera indica* have multiple uses.

Among the total number of 326 plants, 277 are dicots, 42 monocots, 3 bryophytes, one gymnosperm, two pteridophytes and one fungus. These plants are mostly herbs (157), followed by shrubs (81). There are 61 trees, 24 climbers and 3 mosses. These plants belong to 96 families. Of these, 21 families are more significant than others. They have more than five plants used by the tribals. Asteraceae has a total of 19 useful plants and Fabaceae has 18 useful plants.

Among the plants collected, the following are endemic to the Palni hills. *Leucas vestita* Benth. var. *vestita* (Lamiaceae), *Anaphalis lawii* Gamble (Asteraceae), *Zehneria maysorensis* var. *maysorensis* (Wt. & Arn.) Arn. (Cucurbitaceae), *Strobilanthes kunthiana* (Nees) T. Anderson ex Benth. (Acanthaceae), *Kalanchoe bhidei* T.Cooke (Crassulaceae), *Kalanchoe grandiflora* Wall. ex Wight & Arn.

(Crassulaceae), *Didymocarpus innominatus* Burtt. (Gesneriaceae), *Mahonia leschenaultii* (Wall. ex Wight) Takeda (Berberidaceae), *Bentinkia condapanna* A.Berry (Arecaceae), *Tabernaemontana gamblei* Subram. & Henry ex DC (Apocynaceae) and *Heracleum rigens* var. *rigens* Wall. ex DC. (Apiaceae). Among these plants *Zehneria maysorensis* is endangered and *Strobilanthes kunthiana* is vulnerable. Plates 7.1 to 7.8 depict some of plants used by these tribals for various purposes.

Medicinal Plants

A total of 233 plants are used as medicines. Of this 121 are herbs, 58 are shrubs, 40 are trees and 14 are climbers. 204 are dicots, 26 are monocots, one gymnosperm and two pteridophytes. Ninety-four plants are used internally, 66 externally and 73 internally as well as externally. Leaves, seeds, entire plants, flowers or inflorescence, root, sap or latex, yam, rhizome, root pulp, pith, fruits, stem, young shoots and bark are used to prepare medicines. These are used to cure a total number of 192 diseases. Diseases like diarrhoea, cold, cough, indigestion, tiredness, body pain, scabies, attack by evil spirits, delivery, wounds or sores, head ache, fever, stomach disorders, dysentery, body heat, dental problems are the commonly occuring diseases among these tribals and many different plants are used to cure these diseases.

Besides plant remedies, tribals use other methods to cure diseases. They are:

1. Long cloth, wet with cold water is rolled around the stomach of the affected cattle to reduce body heat and heat stroke.
2. Pure honey is mixed with lime powder and applied externally at the throat area and can also be given inside for throat infection and cough.
3. The excreta of a goat is ground and given inside. This induces vomiting and all the indigested materials in the stomach will come out. This is mainly administered for stomach pain due to indigestion.
4. The excreta of monkey is applied on scabies. The intestine of a monkey is sun-dried, cooked and eaten. This cures paralysis and reduces the risk of heart attacks and the swelling of the hands, legs and joints.

5. The intestine of a porcupine is dried, ground, mixed with fresh cow milk and drunk. The person who has taken this medicine should get involved in some hard physical labour or run for some time in order that the medicine is quickly digested and gets into the blood stream. This medicine is given in case of too much of white discharge for woman.
6. A particular measure of water from cold rice is taken and sliced small onions are added to this water. This mixture is put outside the house in dew for one night. The affected person drinks this on the following day early morning in empty stomach. This treatment is given to reduce body heat.
7. Nests of white ants are ground and pepper water is prepared with that and drunk to cure running nose and severe cold.
8. The mud of the anthill is mixed with water and applied at the affected place. This is to cure sprain in the leg or some defects in the nerves.
9. The urine of cow is applied for scabies.
10. The fat of wild boar is put in a banana and given to cattle. It can also be dried, powdered and given to cattle. This is to cure diarrhoea or dysentery in cattle.
11. Monkey's meat is cooked with small onion, dry ginger, pepper and cumin seeds and that extract is distilled and drunk thrice a day to cure asthma, T.B. and wheezing.

Plants Used for Culture

Tribals believe in the effect of witchcraft, evil eyes and black magic. They also believe that evil spirits exist and cause harm to them. There are some plants used to prevent the effect of all these. They are given inside and/or kept with the affected person and/or kept in the house to ward off evil forces. These plants are *Agave cantala, Datura innoxia, Girardinia diversifolia, Pergularia daemia, Plectranthus amboinicus, Rubia cordifolia, Ruta chalepensis, Sansevieria roxburghiana, Toddalia asiatica* and *Zanthoxylum tetraspermum.*

Flowers and leaves of some plants are used for decorating houses, temples and villages during common celebrations in the villages, auspicious occasions and domestic celebrations. These

Plate 7.1
ETHNOBOTANICAL WEALTH

Plate 7.2
ETHNOBOTANICAL WEALTH

Plate 7.3
ETHNOBOTANICAL WEALTH

Plate 7.4
ETHNOBOTANICAL WEALTH

Plate 7.5
ETHNOBOTANICAL WEALTH

Plate 7.6
ETHNOBOTANICAL WEALTH

Plate 7.7
ETHNOBOTANICAL WEALTH

Plate 7.8
ETHNOBOTANICAL WEALTH

plants are *Aerva lanata, Clausena indica, Mangifera indica and Azadirachta indica.* Young shoots of the following plants are grown in mud pots and offered to the village deity during the village annual festival. These plants are *Lablab purpureus, Pennisetum typhoides, Cajanus cajan* and *Sorghum vulgare.*

The sticks of *Clausena heptaphylla* and *Ficus tsjakela* are used to put up sheds to keep the girls who have attained puberty, in seclusion. The plant *Crotalaria pallida* is kept at the entrance of the house of enemies to make them leave the house and village. The plant *Cryptolepis buchananii* is taken while going for seeking alliance for a boy to get a positive reply from the girl's family. The logs of *Ficus racemosa* are used to put up 'pandals' (sheds) for any auspicious occasion like marriage, village festival etc. It is believed that this plant with sap will have a beneficial effect.

The wood of *Gmelina arborea* and stem of *Pinus roxburghii* are used to make musical instruments like drum and flute respectively. The plant *Leucas vestita* is kept near, while boiling paddy or wheat. It is believed that the boiled grains will remain fresh and fertile and the yield of harvest will increase. The leaf of *Strobilanthes kunthiana* is used to find out whether an eloped couple will live together or not.

Food

A total of 85 plants are used for food by these tribals. Of this 85, thirty-seven plants are used only as food. The edible parts include leaves, vegetables, pith, yam, fruits and seeds. Some herbs are used both as medicines and greens. The leaves of these plants because of their edible nature, add essential vitamins to the body, when eaten. Some of these plants are collected from the wild and some from the places of their work. Some of the cultivated vegetables are taken home with the permission of the estate owners. Some of the tribals also grow these plants and vegetables.

Plants provide daily bread to humankind in one way or the other. This is quite true in the case of Palliyans and Pulayans as their hunger is satisfied by both wild and cultivated plants. It is to be noted that many wild edibles are widely consumed throughout the world by various tribes, though they are less known or not known to urban or civilized population. This may be due to the non-availability or the bitter taste in the leaves of some species.

The leaves of *Amaranthus viridis, Commelina benghalensis, Solanum nigrum, Oxalis corniculata*, seeds of *Cleome gynandra*, pith of *Caryota urens*, fruits of *Gmelina arborea, Ziziphus rugosa, Ficus hispida, Mangifera indica* and *Syzygium cumini* are used for edible purposes by these tribals. The same observation has been made by Amuniddin and Girach (1991);[1] Manandhar (1996)[2] and Vartak and Suryanarayana (1995).[3] Among the green leafy vegetables, *Amaranthus* species are widely eaten by rural population throughout India.[4] Other than wild greens, *Coccinia grandis, Sesbania grandiflora* and *Solanum nigrum* are reported to be consumed as under-utilized conventional vegetables.[5]

Cultivated Plants in the Surveyed Villages

Mountains are those areas that have a local relief sufficient to ensure significant vertical ecological differentiation. Certain physical features characterize mountain areas: a decline in atmospheric pressure and temperature as elevation increases and an altitude related rise in short-wave radiation. Topography brings about local variations in climatic conditions and modifies the soil in the habitat, according to altitude, steepness, exposure to light and wind and direction of the mountain chain. In general, water table tends to be nearer the surface of the soil in the valleys than on hills, where it is deeper owing to rapid percolation of water. Consequently draught conditions occur more often in the hills. In the valleys, in spite of extreme porosity of the soil owing to the proximity of the water table, good plant growth is achieved.[6]

Mountain areas are characterized by some important aspects like inaccessibility, marginality and fragility. Table 7.1 gives a list of crops cultivated in the surveyed villages in the Palni Hills. Figure 7.1 gives the cropping calendar. The tribals cultivate some vegetables in the piece of land they have.

Coffee Cultivation

Most of the tribals are tied down to their daily wage labour in different estates. Since coffee does not give them immediate income compared to daily wage, no adequate inputs are given for the coffee lands owned by the tribals. Coffee yields give income only at the fruiting season. But the tribals need money for the day-to-day management of their domestic needs.

Table 7.1
CULTIVATED PLANTS IN THE SURVEYED VILLAGES

Sl.No.	*Vernacular Name*	*Common Name*	*Botanical Name*	*Family*	*Cultivated Variety*
1.	Keppai	Ragi	*Eleusine coracana*	Poaceae	
2.	**Kothumai**	Wheat	*Triticum aestivum*	Poaceae	
3.	**Malai arusi**	Rice	*Oryza sativa*	Poaceae	Sengary
4.	**Beans**	Beans	*Phaseolus vulgaris*	Fabaceae	Selection – 9
5.	**Vellai Poondu**	Garlic	*Allium sativum*	Alliaceae	Singapore, Mettupalayam
6.	**Vengayam**	Onion	*Allium cepa*	Alliaceae	
7.	**Vurulai**	Potato	*Solanum tuberosum*	Solanaceae	Kufry Jothi, Kufrimuthu
8.	**Carrot**	Carrot	*Daucus carota*	Apiaceae	Zino
9.	**Muttaikosu**	Cabbage	*Brassica oleracea var. capitata*	Brassicaceae	Harirani
10.	**Kambu**	Millet	*Pennisetum americanum*	Poaceae	
11.	**Malli**	Coriander	*Coriandrum sativum*	Apiaceae	
12.	**Vazhai**	Banana	*Musa paradisiaca*	Musaceae	Virupachi
13.	**Samai**	Common millet	*Panicum miliaceum*	Poaceae	
14.	**Kappi**	Coffee	*Coffea arabica*	Rubiaceae	Selection – 795, Kaveri, Keltemore
15.	**Ingi**	Ginger	*Zingiber officinale*	Zingiberaceae	

Most of the coffee lands cultivated by the tribals are not patta lands and they are mostly cleared forest lands. So anytime the forest officials can destroy the cultivation or initiate action (in some places it has already been done), against these tribals. That is why intensive coffee cultivation is not done in these surveyed tribal villages.

Plants Used for Domestic Needs

A total of 31 plants are collected and used for domestic purposes by these tribals. Among the 31 plants, 26 are used only for domestic purposes.

The wood of *Gmelina arborea* is used to make furniture and to build houses. The leaves of *Arunda donax* are used to thatch sheds and houses. The leaves, stems, barks and sometimes the entire plants

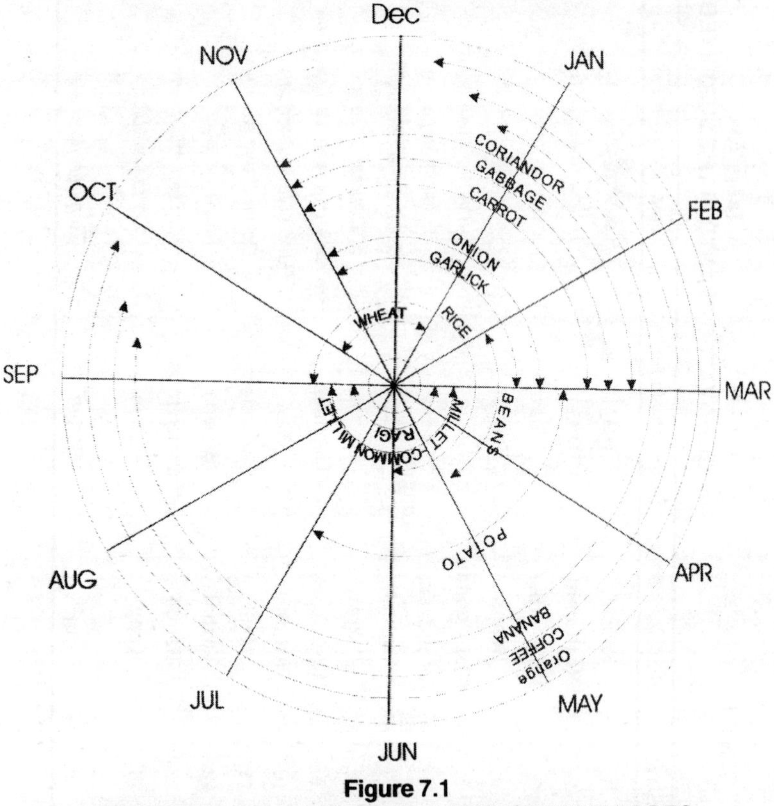

Figure 7.1
CROPPING CALENDER IN THE SURVEYED VILLAGES

of the following are ground and sprayed in the streams and rivers to catch fishes. The fishes faint and float on the surface and they are collected. The plants, used as fish poison are *Tephrosia fusca*, *Hydrocotyle javanica* and *Diospyros nigricans*.

Fresh peeled stem of *Canthium dicoccum* is burnt and used as light to walk at night in the forest. The bark of *Grewia tiliifolia* is soaked in water for one hour and then applied on the hair in the head as shampoo.

The bark of *Osyris arborea* var. *puberula* is used instead of betel nut to chew betel leaves. The wood of *Commiphora caudata* is used to make pounder and pounder base. The branches of *Aristida setacea* are tied together and used as broom. The following grasses are used by these tribals to thatch houses: *Themeda cymbaria*, *Themeda tremula*, *Themeda triandra*, *Cymbopogon martinii*, *Cymbopogon nardus* var. *confertiflorus* and *Cymbopogon travancorensis*. The leaves of *Euphorbia rosea*, *Ziziphus oenoplia* are used to stall-feed the goats.

Firewood and Cooking

Firewood remains the major source of domestic fuel in hilly regions. Easy access for collection from nearby forests has not allowed the villagers to resort to the use of some other fuel for domestic consumption.[7] Firewood is collected from Orange, Jack, Wattle and Erythrina trees in the coffee estates. In case the estate owners do not permit these people to collect firewood from the estates, then it is collected from the forests nearby. Generally the tribals, both men and women, get involved in firewood collection on Sundays. The species collected for firewood are *Mahonia leschenaultii*, *Ixora nigricans*, *Ailanthus excelsa*, *Grevillea robusta*, *Citrus aurantium*, *Artocarpus heterophyllus*, *Coffea arabica*, *Coffea robusta*, *Ficus benjamina*, *Polygala arillata*, *Toona ciliata* and *Melia azedarach*.

Cooking is done over an open fire. So a large scale of dispersion of flame during windy conditions and a lack of proper control over the fire, lead to excessive loss of energy. However, this practice is useful for space-heating during cold climate. Cooking is done in the morning and evening.

Animal Hunted by the Tribals for Food

The tribals are good in hunting animals and they trap mouse deer, boar, wild fowl and flying squirrels. A good catch is a great boon for the famished tribals. The meat obtained is divided among all families within the settlement. The animals hunted are given in Table 7.2. Wild pigs and mouse deer of any size are the two animals that are often hunted. The tribals use leaves and bark of some plants as baits. Fish poisons are used in still water, while in running water, fishes are caught using cloth nets, either by sweeping the cloth through the water or by making a small stone dam which spills through the pole-propped cloth.

In some places monitor lizard, mouse deer, porcupine, and wild cat are hunted with the help of dogs. The dogs chase these animals. On seeing them, in order to escape, they hide themselves in their respective dwelling places (holes). Then they burn some dry sticks and let the smoke into the hole. Because of suffocation and heat, the animals die and in that case the hole is dug and the animal is removed. In some cases, when the smoke is let in, the animals try to escape and at that time they are caught. The following other methods are used to catch other animals:

1. Flying squirrel is hit with stones or sticks and killed.
2. Wild boars are caught with special wire nets made for this purpose or they are killed when biting liver-coated country made bombs that are put in the corridors of these animals.
3. Nilgiri tahr is shot at and caught. The tribals accompany the estate owners who have guns. They all go for hunting and get a share in the hunted animals.
4. Fishes are caught with dhothies used as nets.
5. Small fishes are tied to the tip of a small stick and the stick is lowered into a crab hole. When the crabs try to catch and eat the fishes, the stick is slowly taken up and the crabs are caught by hand.

The hunted animals are cooked and eaten.

Non-Timber Forest Products (NTFPs) Collected by the Tribals

Non-Timber Forest Products refer to all biological materials other than timber which are extracted from natural forests for human use. These include food, medicines, spices, essential oils, resins, gums, latex, tannins, dyes, ornamental plants, wild life (products and live

Table 7.2
ANINALS HUNTED FOR FOOD BY THE TRIBALS

Sl.No.	Vernacular Name	English Name	Zoological Name	Phylum	Sub-Phylum	Class
1.	Kooran/Saruguman	Indian chevrotain or Mouse deer	*Tragulus meminna*	Chordata	Vertebrata	Mammalia
2.	Kattu Eli	Field Mouse	*Mus booduga*	Chordata	Vertebrata	Mammalia
3.	Udumbu	Monitor	*Varanus bengalensis*	Chordata	Vertebrata	Reptilia
4.	Kelai adu/Varai adu	Nilgiri tahr	*Hemitragus hylocrius*	Chordata	Vertebrata	Mammalia
5.	Mullam pandri	Porcupine	*Hystrix indica*	Chordata	Vertebrata	Mammalia
6.	Keeri	Mungoose	*Herpestes edwardsi*	Chordata	Vertebrata	Mammalia
7.	Kattu koli	Greg jungle fowl	*Gallus sonneratti*	Chordata	Vertebrata	Aves
8.	Kili	Parrot	*Psittacula krameri*	Chordata	Vertebrata	Aves
9.	Kuruvi	House sparrow	*Passer domesticus*	Chordata	Vertebrata	Aves
10.	Kurangu	Monkey	*Macaca radiata*	Chordata	Vertebrata	Mammalia
11.	Nandu	Stream Crab	*Paratelphusa hydrodromous*	Arthopada	Invertebrata	Crustacea
12..	Kattu pandri	Wild pig or Wild boar	*Sus cristatus*	Chordata	Vertebrata	Mammalia
13.	Meen	Fish	*Mugilocur* spp.	Chordata	Vertebrata	Pisces
14.	Aamai	Tortoise	*Testudo elegans*	Chordata	Vertebrata	Reptilia
15.	Alungu	Pengolin	*Manis crassicaudata*	Chordata	Vertebrata	Mammalia
16.	Verugu	Jungle cat	*Felis chaus*	Chordata	Vertebrata	Mammalia
17.	Vavaal	Flying Fox	*Pteropus gigantens*	Chordata	Vertebrata	Mammalia
18.	Paravai anil	Flying squirril	*Petaurista petaurista*	Chordata	Vertebrata	Mammalia
19.	Karaiyan	Termite	*Odontotormes* sp.	Arthropoda	Invertebrata	Insecta

animals), firewood and materials notably rattan, bamboo, small wood and fibres. The tropical moist forests have incredibly large number of such products exploited by millions of people. These NTFPs have always been critical to meeting the basic needs of the rural population and have been major commercial products since they first entered international trade many centuries ago. There are about 29 million forest dwellers most of whom are tribals in Southeast Asia who are dependent on these resources.[8]

The functions fulfilled by NTFPs in the rural household economy include providing: food security in the form of staples or more frequently nutritional supplements, such as snacks and side-dishes and buffers against seasonal and emergency shortages; health care in the form of ingredients in traditional medicines; raw materials for building and of implements for household use or uses in support of their economic activities like agriculture, fishing, hunting or small scale processing and manufacturing enterprises; and finally income and employment from the collection, trade or processing and manufacture of NTFPs.[9] The forest products are significant component of tribal diet[10] and collectively they provide protein, energy, fibre, vitamins and essential minerals to the human diet.[11]

The role of NTFPs in the national economy involves providing income and materials for direct consumption or for use in industries, large and small; employment in collection, trade, processing and manufacture and much needed foreign exchange through export of raw materials and processed or finished products.[12]

The following is a brief review of some studies done outside India on NTFPs:

A study done in Southeast Asia with special reference to Indonesia, Malaysia and Thailand by Jenne and Malani (1989)[13] revealed that the following products are collected from the forests by the tribals:

1. Medicines
2. Fodder, fuel wood and charcoal
3. Structural materials such as rattan, bamboo, wood poles and various fibres
4. Chemically useful materials including resins, essential oils, gums, latexes, tannins and dyes

5. Animal products such as honey, eggs, bird nests, lac, silk, reptile skin, feathers and other decorative wildlife products
6. Live animals–they hunt hoofed mammals, the ungulects, wild boar, deer, long-tailed or crab eating macaque, ant-eaters and bats
7. Ornamental plants.

This study further concluded that

1. The economic value of NTFP in Southeast Asia is highly significant to the rural and national economies of the region in terms of the provision of material needs, cash income and employment.
2. The economic importance of these resources is greater for the rural population of the forest than for the others. So it should be managed and 'developed'.
3. NTFP extraction can represent a sustainable form of tropical forest exploitation, for which traditional forest management systems provide valuable models.
4. There is not only greater economic potential inherent in conserving and enhancing the forest resource base while developing the marketing and local processing of NTFP, but a crying need to do so. If no action is taken, the way of traditional forest dwellers will disappear and with it will be lost the share of knowledge they hold regarding the use and sustainable management of non-timber forest resources.

Hollowes and Alexander (1990)[14] found out that forty-eight species or groups of native plants are used as medicines and as food or beverages by the Gevich in Athabaskan and Caucasian residents. Most widely accepted medicine in the Fort Yudon area is spruce (*Picea glauca*) pith. Wild fruits and mushrooms comprise the greatest proportion of edible plants. Most common fruits harvested are *Viburnum edule* and *Rosa acicularis*.

The Chumash Indians occupying the mainland and offshore islands in the vicinity of Santa Barbara, California, use atleast 150 plant species for food, medicine, material culture and religious practices.[15]

The people of Kwara state, Central Nigeria, use 17 plants as food, 3 as cosmetics, one as insecticide and the other for commercial purposes. On the whole they use 52 plant species for different purposes.[16]

The Miskutoes make use of 23 plants to cure snakebite and athlete's foot. The other commonly used plants are *Desmodium incanum*, as antiseptic for wounds, *Acranthera grandiflora* to counteract skin infections, *Senna occidentalis* for pain in general and *Sida acuta* to speed up deliveries.[17]

As many as 95 plant species, both wild and cultivated, representing 88 genera under 47 families have been found to be of common use for the treatment of 23 types of diseases among the Tamangs, one of the ethnic tribes of Nepal.[18]

Coming to the Indian scene, ethnobotanical researchers have so far brought on record over 500 plants as significantly used by the tribals as food, dyes, tanins, drugs, narcotics, drinks, housing instruments, weapons, fibres and medicine.[19] But broadly speaking about 800 plant species are used as food alone by the tribal inhabitants.[20] Yet another report mentions that one third of the higher plants in India are used by the tribals and poor people for food, medicine etc.[21]

A variety of plant food products are brought by the tribals for sale in the local markets for consumption by the urban population as for example, the fruits of *Elaeocarpus floribundus*, *Docynia indica*, *Prunus jenkinsii* in north-eastern region, *Rhodomyrtus parviflora* in Nilgiris, *Myrica nagi*, *Rubus ellipticus* and *R. lasiocarpus* in the Himalayas, *Grewia populifolea*, *Zizyphus*, *Cordia rhus* and *Salvadora* species in the drier tracts and the tubers or rhizomes of *Dioscorea*, *Colocasia*, *Nymphaea*, leaves of *Colocasia*, *Ipomoea aquatica*, flower buds of *Bauhinia* and the kernels of *Buchanania lanzan* (chironji) in most parts of penisular India.[22]

Coming to certain specific studies done on certain specific tribal groups, Arora (1981)[23] found out that the tribals in the northeastern region use about 300 plant types (tubers, greens, flower buds, ripe fruits) for food. Studies by Joseph and Knarkongon (1981)[24] on the tribals of Khasi and Hainatia hills in Meghalaya revealed that these people use over one hundred plants for various purposes such as medicine, subsidiary food (*Flemingia vestita*, *Gynocardia odorata*,

Houttuyina cordata), making implements (*Quercus dealbiata, Camellia caduca*) and musical instruments, religious ceremonies (*Lagenaria siceraria, Zingiber officinale*).

The Miris or Mishings, a tribe living in the Brahmaputra valley of Assam, use thirty-two plants for medicine, twenty-eight for local drinks, twenty-five for construction purposes and fishing implements, twenty-three for fruits and vegetables, two as substitute for soap and two for rearing silk worms for thread.[25]

The Karbis, a hill tribe of Assam inhabiting the mountainous regions use wild plants for:

1. Subsidiary food, *e.g.* kernel of *Entada phaseoloides*, bark of *Premna latifolia*, base of rachis of *Angiopterivs erecta* etc.
2. Thatching: Leaves of *Calamus erectus, C. latifolius, Daemonorops jenkinsianus* etc.
3. Building materials in house building: Barks of *Sterculia villosa, Trema orientalis*.
4. Medicine: Powder of dry fruits of *Brucea mollis* in malarial fever, leaf-juice of *Clausena excavata* in musular pain, root decoction of *Ixora cuminata* as galactagogue.[26]

Over hundred and twenty flowering plants are known for their use as food by the tribals residing along the hilly regions of the Western Ghats in Maharastra state and the state of Goa. Of these, plants fifty-eight are commonly used, twenty-seven are occasionally used and the remaining are reluctantly consumed as food during periods of severe drought and extreme scarcity of other conventional foods.[27]

The Nilgiris is the home of Todas, Kotas, Irulas, Kurumbas, Paniyas and Kattunayakas. The Todas who are the earliest known inhabitants of the Nilgiris and are pastoral in nature use thirteen plants for medicine and eleven for food. The Kotas who are artisans and cultivators use twenty plants for medicine and fourteen for food. The Irulas who are food gatherers, primitive agriculturists and hunters use eleven plants for food.[28]

The Bhotias are a nomadic and semi-nomadic tribe of Kumaun Himalaya who build huts with bamboos and thatch it with *Chrysopogon gryllus*. They make agricultural implements from the woods of *Quercus incana, Q. glauca, Q. semecarpifolia, Carpinus viminea*,

Toona ciliata, Terminalia alata. Spinning and weaving is the main occupation of tribes in Johar valley. They produce woollen goods of high quality, *e.g.*, shawl (*Pankhi, Pashmina*), quilt (*Chutha, Thulma*) and carpets (*Dan, Ashan*).[29]

The following NTFPs are collected and marketed by the tribals in Orissa–tamarind, tamarind seeds, myrobalans, gooseberries, sal resins, *Rauvolfia serpentina* roots, *Kiddakaya, mahua* seeds, *karanjia* and *kusum* seeds, neem seeds, arrow roots and lac. There are eighty-three plants identified as used by these tribals for food[30] Mango, kendu, jackfruit, tamarind, gooseberries, sitaphal etc. are some of the important fruits consumed by these people. Mahua flower is their main food in lean season.[31] Forty-six wild plants are used for several common ailments by the Pauri Bhuirya of Orissa and main items of NTFPs collected, consumed and sold are sal seed, mahua flowers, Kusum seeds, char seeds, amla and honey.[32]

Singh and Maheswari (1990)[33] found that the Tharus of Nainital district in Utter Pradesh make use of two hundred plants and plant produce for food, fibre, medicine, oils, handicrafts, fuels etc.

Study of Maheswari (1990)[34] on the Baigas of Madhya Pradesh revealed that these people use seven hundred and eighty two plant species belonging to seven hundred and nineteen genera and four hundred and thirty families.

A total of 171 plants species are used by the Nishis, the Hill Miris, the Sullungs and the Apatains of the Lower Subansiri district of Arunachal Pradesh in northeastern India. Twelve animal species are used by the Nishis. Of the total plant species recorded, 38 per cent are leafy vegetables and 28 per cent are edible fruits. Others are used for food, liquor, medicine, household goods, traditional dresses, dyes, tattoos, fish and animals poisons or for salt extraction. The Sulungs use starch obtained from the pith of *Metroxylon segu* as a staple food.[35]

A study by Malhotra, Poffenberrger, Bhattacharya and Dev (1991)[36] in West Midnapore district in Southwest Bengal, recorded the names of 29 NTFPs available in a forest that have been protected for three years.

Madhya Pradesh has some of the largest and richest forests in India and accounts for 26.3 per cent of the total forest area in the country. A large number of herbs, crude drugs and plant products are collected from the forests and are categorized as NTFPs.[37]

Of the total wage employment in the forestry sector, NTFPs have accounted for more than seventy percent.[38] Gupta and Guleria (1982b)[39] and ISST (1987)[40] have estimated the production and employment potential of NTFPs in India.

1. Production of oil seeds from trees (Mahua, Neem, Karanj, Kusum and sal are the main species) in 1981-82 was 0.5 million tonnes. Oilseeds from forests could perhaps substitute for part at least of the over 1.5 million tonnes edible oil (equivalent to 12 million tonnes of oil seeds) imported annually.
2. *Tasar silk*: This commodity is produced from sal, arjun, and oak trees by tribals spread over seven states. India produces 22 per cent of the total world production and about one lakh tribals are engaged in this.
3. *Tamarind*: This fruit is collected during December to March. A good tree yields 1.5 quintals of tamarind fruit annually, valued at Rs.600. The annual marketed production of tamarind fruit is about 2,50,000 tonnes.
4. *Grasses and leaf fodder*: These are the other important NTFPs obtained. The committee on Fodder and Grasses estimated that in 1985 only about 57 per cent of dry fodder and 27 per cent of green fodder requirements were being met.[41]

Seventy per cent of NTFPs collected are from the five states of Maharashtra, Madhya Pradesh, Bihar, Orissa and Andhra Pradesh where 65 per cent of the tribal population live.[42] Thirty-five per cent of the earning of the tribals in a district of Gujarat and up to thirty-eight per cent in Madhya Pradesh have been found to come from these NTFPs.[43]

Though these NTFPs, collected by the tribals contribute greatly to the domestic and national income, the tribals are exploited by the middlemen, contractors and government officials who employ them in the collection of these products.

There were three notable changes that took place in the NTFP sector recently:[44]

1. A decline in NTFP resources has been brought about by widespread deforestation and forest degradation
2. Commercialization of the rural economy has resulted in the increased penetration of the market economy into all aspects of rural life
3. Institutional changes have been imposed by Governments affecting forest land classification and associated regulations, land tenure arrangements, hunting and collecting permit and royalty systems, trade and industry subsidies and regulations and developmental schemes.

All these change led to a three-sided exploitation of the tribals:

1. Low price given for the products collected
2. Lending money in lean months in order to get NTFPs at very low prices in the harvesting season
3. Cheating in measurement.

There is no other way of escape from these forces because of the monopoly of the contractors and the tribal dependence on the income.

The tribal people are continuously losing command over their resources on almost all counts. The states are levying royalties on the NTFP and the tribals are getting just the wage for collection (28th Report of the Commissioner on SCs. and STs., 1986-87).

The sons and daughters of forests are economically, ecologically and culturally destabilized. Collection, processing and sale of NTFPs have been the traditional backbone of their existence. They usually sell the products through various agencies where they do not get a fair price. Chiefly middlemen make advance purchase of the products at deplorably low rates. Hence they are subjugated to a double-sided agony: depletion of the resources on the one side and low sale value for the same on the other.[45]

In many states so-called NTFP has become a major source of forest revenue. Till recently except for a few items, others had no market value and the tribals had full access to the same for generations. Now the state has declared monopoly right over most of these items and the tribals sell such non-timber forest products to State Forest Corporations or their purchasing agents. As the state is considered owners of the products, the payments that the forest dwellers receive are really labour charges for collection and not prices for the commodities placed by them in the market. Some studies show that the hill dwellers get only a fraction-sometimes as low as ten per cent, of the price at which the consumers buy them through the state sponsored or state supported channels.[46]

Products are collected and sold to businessmen or Government Corporation in Madhya Pradesh and Orissa. The price which they get for tamarind is as low as Rs.1.40 to 1.75 per kg. whereas it sells for Rs.9 to 11 per kg. in the Bombay market.[47] In 50 out of 68 villages of Orissa, it was found[48] that Government agencies had not managed to eliminate middlemen. The same middlemen who till recently exploited the tribals as moneylenders and merchants continue their work in the garb of agents of Government bodies. After meeting all the expenses and profits, the tribals should be paid Rs.1.30 a kg for collection of sal seeds in Madhya Pradesh but the Government pays only 55 paise a kg.[49] By the sale of leaves in Orissa, the suppliers get Rs.19.79 whereas the petty businessmen get Rs.222.50. The weaker section get relatively low price per unit while the merchants get much higher income from the same unit of a produce.[50]

Some tribal regions have no regular markets. Private wholesale dealers go to the 'mandis' which are held once a week to buy the NTFP at very low rates and sell them at huge profits in the big cities. Many a time they make an advance payment to the tribals and later buy their goods at nominal costs.[51] Most of the tribals do not know to count, and the price that is mentioned often does not correspond to the amount they get in hand. Generally the trader is a moneylender too who buys the NTFPs from the tribals as a repayment of the debt. But the price fixed is incredibly low.[52]

Besides getting a low price for the products they collect, the distance travelled for collection has also increased drastically due to deforestation.

The examples of Orissa and Chattisgarh reveal that owing to deforestation, an average of 7 km has to be covered for NTFP collection now as compared to 1.7 km about 20 years ago.[53]

Studies by Menon and Geeta (1987, 1991),[54] Dasgupta Supriya (1988)[55] and Ali (1980)[56] also confirm the hardships experienced by tribals especially tribal women due to deforestation and other exploitations.

Several NTFPs were nationalized in the interest of revenue maximization which seemed to have helped contractors and petty officials to the disadvantage of the people who depended upon them.[57] Nationalization has also brought down the production. For example the production of Tendu leaves in Madhya Predesh fell from 5.10 million bags in 1981-82 to 3.9 million bags in 1985-86.[58] The production of sal seed in India fell from two lakh tonnes in 1979 to only 60,000 tonnes in 1987. Lac, another nationalized item, has gone down from an average of 32,000 tonnes per year during 1961-70 to 16,000 tonnes during 1981-86.[59]

NTFPs in the Palni Hills

A total of 30 products are being collected as commercial non-timber forest products by the tribals in the Palni hills. Among these, 29 are collected from plants. Collection of these products remains the principal subsidiary occupation for these tribals. The details of the products collected are given in Table 7.3. Moss, gallnut, gooseberry, *Passiflora* and date-palm are available in plenty and they are mostly found in Kodaikanal forest division. Generally, all these products are available in all the forest divisions. Males are involved more in the collection than females. This may be due to the hard labour involved in the collection and the long distance travelled for the collection. Collection of moss demands the ability to climb big trees. This is mainly done by youngsters. Women collect moss from rocks and fallen twigs and branches.

Only very few are involved in NTFP collection as the principal occupation because of the following reasons:

1. Due of deforestation, the tribals have to walk a long distance to collect NTFPs and so only healthy people are involved in this.
2. Generally, only young people are involved in this because of the hard labour involved and sometimes it is also risky (like climbing the trees).
3. In spite of the hard labour involved in the collection of NTFPs, the price paid for the same is very low.
4. Since the availability of NTFPs is seasonally conditioned, continuous employment cannot be ensured.

Table 7.3
DETAILS OF NTFPS COLLECTED BY THE TRIBALS IN THE PALNI HILLS

Sl.No.	*Vernacular Name Tamil (English)*	*Botanical Name*	*Family*	*Parts Collected*	*Seasonal Availability*
1.	Pasam (Rock moss)	*Politricum densifolium*	Bryophyta (Division)	All parts	All months
2.	Pasam (Bog moss)	*Funeria hygrometrica*	Bryophyta (Division)	All parts	All months
3.	Pasam (Tree moss)	*Barbula convoluta*	Bryophyta (Division)	All parts	All months
4.	Kadukkai (Gallnut)	*Terminalia chebula*	Combretaceae	Fruits	Dec-May
5.	Nelli (Gooseberry)	*Phyllanthus emblica*	Euphorbiaceae	Fruits	Oct-Jan
6.	Thurai pudalai (Passion fruit)	*Passiflora edulis*	Passifloraceae	All parts	Jan-March
7.	Injam (Date-palm)	*Pheonix sylvestris*	Palmae (Arecaceae)	Leaves	All months
8.	Injam (Date-palm)	*Pheonix loureirii*	Palmae (Arecaceae)	Leaves	All months
9.	Indu Kodi	*Acacia caesia*	Mimosaceae	Bark	Jan-May
10.	Kattu Milagu (Wild pepper)	*Piper nigrum*	Piperaceae	Seeds	Oct-Nov
11.	Mangai (Mango)	*Mangifera indica*	Anacardiaceae	Fruits	April-May
12.	Sigakai maram (Soapnut)	*Acacia sinuata*	Mimosaceae	Fruits	April-May
13.	Tellikkai	*Entada pursaetha*	Mimosaceae	Seeds	Mar-June
14.	Senbagam (Chempak)	*Michelia champaca*	Magnoliaceae	Flowers	June-July
15.	Karuppu kungiliam (Black dammar)	*Canarium strictum*	Burseraceae	Seeds	Oct-Nov

Contd...

Table 7.3–Contd...

Sl.No.	*Vernacular Name Tamil (English)*	*Botanical Name*	*Family*	*Parts Collected*	*Seasonal Availability*
16.	Kalakkai	*Carissa carandas*	Apocynaceae	Fruits	April-May
17.	Malai Shenbagam	*Michelia nilagirica*	Magnoliaceae	Flowers	April-May
18.	Poosakkai	*Sapindus emarginatus*	Sapindaceae	Fruits	March-May
19.	Vellamarudu	*Terminalia arjuna*	Combretaceae	Fruits	April-May
20.	Vengai maram (Indian kino tree)	*Pterocarpus marsupium*	Fabaceae	Fruits	Oct-Jan
21.	Kochan kodi (Balloon vine or blister creeper)	*Cardiospermum canescens*	Sapindaceae	Stem and Entire plant	All seasons
22.	Thailapul (Lemon grass)	*Cymbopogon flexuosus*	Poaceae	Leaves	All seasons
23.	Pepper (Milagu)	*Piper mullesua*	Piperaceae	Seeds	April-May
24.	Thatpoot kodi	*Passiflora calcarata*	Passifloraceae	Entire plant	All seasons
25.	Nei kottan maram	*Lepisanthes tetraphylla*	Sapindaceae	Fruits	March-April
26.	Samburani maram (Benzoin tree)	*Styrax benzoin*	Styraceae	Sap	All seasons
27.	Elam (Wild cardamomum)	*Elettaria cardamomum*	Zingiberaceae	Seeds	January-March
28.	Kattu chiragam (Wild cumin seeds)	*Vernonia anthelmintica*	Asteraceae	Seeds	Oct-Nov
29.	Koonthapanai (Dwarf palm tree)	*Caryota urens*	Arecaceae	Leaves	All seasons
30.	Then (Honey and wax)				March-May

Because of the seasonal factor, many are involved in NTFP collection only in peak seasons of the different products. That is why the main subsidiary occupation of the tribals is NTFP collection (60 per cent).

Commercial NTFPs are those that are leased out by the Forest Department (F.D.) and the tribals are involved in the collection and sale of these products. Some non-leased products are also collected by these tribals.

The F.D. has divided the reserve forests in the Kodaikanal Forest Divisions into five ranges and leases out these ranges to the highest bidding contractors for the purpose of NTFP collection. The five ranges where these reserved forests are existing are Kodainal range, Perumpallam range, Mannavanur range, Poombarai range and Periakulam range. From 1970–71 upto now, the revenue from the gallnut leasing remains the major source of income for the F.D. The annual expected yield is different for the same product in different ranges because of the difference in the yields of these products in the different ranges. The details of the reserved forests are given in Appendix 4. The yield depends on:

1. Number of trees of that particular product.
2. Their yield potential–saplings do not yield as much as the well grown trees.

The F.D. fair price for the same product differs from one range to another because of the differences in expenditure for collection and transportation. The F.D. fair price for each product is calculated by multiplying the expected annual yield with the F.D. fixed fair price per kg. for each product.

Some of the products like moss, date-palm, benzoin etc., are available throughout the year and this has the potential for round the year employment. Other products are seasonal. To collect gallnut and gooseberry, the males climb the trees and shake the trees or cut the edges of fruit bearing branches. If females are accompanying them, they do the collection of all the fruits fallen on the ground. In order to harvest the products quickly, some times the fruit bearing branches are cut and the fruits are removed. This practice affects the yield of the products in the subsequent years.

These tribals are experts in collecting honey. To get honey from the top of precipitous rocks, they climb with pot and torch in hand, on twisted creepers to the level of the hive, their bodies being entirely unprotected. The find of a beehive in the hollow of some tree is a vestibule feast for them. They smoke the bees out and collect the combs. The tribals from four to seventy years old are agile on the trees. Large, unbranching trunks may, however, require vine-tied ladders of wood. The blooming of *Stenosiphonium russellianum* indicates the season of honey collection. For honey collection, they always go with their brother-in-law or alternatively father-in-law and son-in-law and never do this with a brother.

The collected products are sold to contractors. The price paid for the products to the tribals is very low compared to the range of profit they make out of this transaction. The details of the percentage of profit are given in Table 7.4. So the actual beneficiaries from the NTFP trade are the middlemen and contractors.

A comparative account of the NTFPs collected by different tribal groups is given in Table 7.5. Plates 7.9 and 7.10 bring out certain aspects related to NTFP.

Economic Importance of the Non-timber Forest Products

Moss (*Fumeria hygrometrica*)

1. Because of its water absorbing capacity, these plants are used as packing material for transporting living plants and as a water retaining constituent for preparing seed beds.
2. It serves as a raw material for the production of many industrial products such as acetic, humic and other acids, alcohols, paraffin, naphtha, lignin, etc. This is also used for making paper, polyphone and woven fabrics and as a medium for bacterial culture.
3. In the European countries, it is used as packing material and for sound proofing the wooden houses.

Gallnut (*Terminalia chebula* Rebk.)

1. Myrobalan is one of the principal bloom-yielding tans due to its high ellagitannic acid content, and is useful especially in the production of sole-leather. It helps in the

fermentation of the tan liquor and consequent pulping of the leathers. Its roots, bark and wood also contain tannin.

Table 7.4
PERCENTAGE OF PROFIT FOR THE CONTRACTORS AND MIDDLEMEN FROM NTFPs

Sl.No.	*Names*	*Unit*	*Buying Price (Rs./Unit)*	*Selling Price Price (Rs./Unit)*	*Profit Percent-age*
1,2,3	Moss	Kg	25.00	75.00	200.00
4.	Gallnut	Tin (about 10 kg)	35.00	85.00	142.86
5.	Gooseberry	Kg	5.50	15.00	172.73
6.	Passion fruit	Kg	6.00	24.00	300.00
7,8	Date-palm	Bunch	1.50	6.00	300.00
9.	Indu kodi	Kg	4.00	12.00	200.00
10.	Kattu milagu	Kg	5.00	25.00	400.00
11.	Mango	Kg	1.20	11.20	833.33
12.	Soapnut	Kg	3.60	30.00	733.33
13.	Tellikkai	Kg	1.80	12.00	566.67
14.	Chempak	Kg	7.80	30.00	284.62
15.	Black dammar	Kg	4.00	60.00	1400.00
16.	Kalakkay	Kg	6.00	17.00	183.33
17.	Malai shenbagam	Kg	8.00	32.00	300.00
18.	Poosakkai	kg	5.00	10.00	100.00
19.	Vellamarudu	kg	8.00	12.00	50.00
20.	Vengai maram	Kg	2.00	6.00	200.00
21.	Kochan kodi	Kg	15.00	30.00	100.00
22.	Lemon grass oil	Kg	200.00	350.00	75.00
23.	Milagu	kg	5.00	25.00	400.00
24.	Thatpoot kodi	kg	15.00	30.00	100.00
25.	Nei kottan maram	Tin (about 5 kg)	50.00	200.00	300.00
26.	Samburani maram	kg	40.00	230.00	475.00
27.	Elam	kg	300.00	500.00	66.67
28.	Kattu chiragam	kg	60.00	120.00	100.00
29.	Koontha panai	One measure (about 50 leaves)	175.00	250.00	42.86
30.	Honey	bottle (750 ml.)	60.00	120.00	100.00

Table 7.5
DIFFERENT NTFPs COLLECTED BY THE TRIBALS IN DIFFERENT PLACES A COMPARATIVE ACCOUNT

Sl.No.	Tribals	Place	Plants Collected	Reference
1.	Gevich	Alaska	48	Hollowes and Alexander, 1990[60]
2.	Chumash Indians	California	150	Timbrook, 1990[61]
3.	In Kwara state	Central Nigeria	52	Bhat *et al.*, 1990[62]
4.	Miskutoes	Eastern Nicaragua	23	Dennis, 1988[63]
5.	Tamag	Napal	95	Manandhar, 1991[64]
6.	Northeastern tribals	India	300	Arora, 1981[65]
7.	Tribes of Khasi and Hainatia hills	Meghalaya	100	Joseph and Kharkongor, 1981[66]
8.	Miris	Brahmaputra valley	112	Baishya and Hajra, 1981[67]
9.	Tribes in Western Ghats	Maharashtra and Goa	120	Vartak, 1981[68]
10.	Todas	Nilgiris	24	Abraham, 1981[69]
11.	Kotas	Nilgiris	34	Abraham, 1981[70]
12.	Irulas	Nilgiris	11	Abraham, 1981[71]
13.	Pauri Bhuirya	Orissa	52	Mohapatra, 1991[72]
14.	Tharus	Utter Pradesh	200	Singh and Maheshwari, 1990[73]
15.	Baigas	Madhya Pradesh	782	Maheshwari, 1990[74]
16.	Nishris, Hill Miris, Shullungs and Apatanis	Arunachal Pradesh	171	Gnagwar and Ramakrishnan, 1990[75]
17.	Palliyans and Pulayans	Kodaikanal Hills	30	Present study

Plate 7.9
NTFP IN PHOTOS
A: Youngsters getting ready for moss collection; B: Climbing the tree for moss collection; C: Moss being collected from a fallen twig; D: Collected moss being sun-dried; E: Moss for sale; F: Phoenix leaves after collection

PLATE 7.10

G: *Passiflora calcarata* in the wild; H: Gallnut being sun-dried; I: A simple device for lemon grass oil extraction; J: *Pterocarpus marsupium*–full of fruits awaiting a tribal hand K: Tribals on a *Michelia nilagirica* tree to collect the flowers; L: Honey-comb for sale; M: Implements used for NTFP Collection

2. The extract is used on an extensive scale for the internal treatment of locomotive feed-waters and as an additive in oil-drilling composition.
3. The Myrobalans give an yellow dye with alum, and a good black ink with a warm solution of iron-sulphate. It is also mixed with mortar in house building; pericarp is also used as a purgative.
4. Fruits are used externally as local application to chronic ulcers, wounds and as dentarfice and ulceration of gums.
5. Coarsely powdered fruit is smoked in asthma.
6. The bark is diuretic and cardiotonic. The kernels yield a fatty oil. Trees yield gum.

Gooseberry (*Phyllanthus emblica* L.)

1. The fruits are probably the richest known natural source of vitamin C. They are sour and astringent, cooling, diuretic, laxative.
2. The bark and fruit are used in tanning as they give a blackish dye. (tannin content is 28 per cent in fruits, 8-21 per cent in bark and 22 per cent in leaves).
3. The fruits are used in the preparation of writing inks and hair dyes.
4. The dried fruit is detergent and is used as shampoo for the head.
5. The seeds are used in the treatment of asthma, bronchitis and biliousness.
6. The leaves and fruits are used as fodder for cattle.

Passion Fruit (*Passiflora edulis* Sims.)

1. The juice is highly acidic and is preserved for use in blends with less acidic fruit juice and in the preparation of squashes, cordials, syrups, carbonated beverages, jellies, etc. It is also used for flavouring candy, ice-creams, cake-fillings and frostings.
2. Peels can be used for recovery of pectin, as stock feed and as manure.
3. Seeds yield a semi-drying oil, suitable for use in plants and varnishes.

Date Palm [*Phoenix sylvestris* (L.) Roxb.]

1. The fruits are sweet and edible
2. The pith from the upper portion of the stem is eaten
3. The leaves are used to make mats, baskets, fancy handbags, brooms, ropes, etc. and they are also suitable for cordage. The leaves are also used for thatching.

Indu Kodi [*Acacia caesia* (L.) Willd.]

1. The bark is used as a substitute for soap for washing hair
2. Bark extract is employed as a protective coating for boats and fishing nets

Kattu Milagu (*Pipper nigrum* L.)

1. Fresh green pepper is sometimes used for preparing pickles
2. Black and white pepper are among the major condiments employed for seasoning freshly cooked and prepared foods
3. It is used to preserve meat
4. Whole fruits are added to pickles, certain types of sausages etc.
5. In modern Indian medicine, it is much employed as an aromatic stimulant in cholera, weakness following fever, vertigo, coma etc.as a stomachic in dyspepsia and flatulence and as an antiperiodic in malarial fever and as an alternative paraplegia and arthritic diseases
6. For local application for relaxed sore throat, piles and some skin diseases

Mango (*Mangifera indica* L.)

1. Mango is the most popular and choicest fruit of India and occupies a prominent place among the best fruits of the world. It has a rich, luscious, aromatic flavour and a delicious taste in which sweetness and acidity are delightfully blended. Young and unripe fruits are usually acidic and used in pickles, chutney, amchur and culinary preparations.

Soapnut [*Acacia sinuata* (Lour). Merr.]

1. The pods are extensively used as a detergent and the dry ones are regularly sold in bazaars. In recent years, the powdered pods, sometimes perfumed are also marketed. This is preferred to soap when taking oil bath as it does not leave the skin dry
2. Tender acidic leaves are used in chutneys
3. The pods are also used as fish poison
4. Decoction of pods are purgative and relieves biliousness
5. Bark is used for tanning fishing lines.

Thellikkai (*Entada pursaetha* DC.)

1. The seeds are considered tonic, emetic, antiperiodic and anthelmintic

Malai Shenbagam (*Michelia nilagiriea* Zenk.)

1. Champak flowers are employed in India in the preparation of attars and perfumed hair oils. Their oil constitutes one of the most exquisite raw materials for perfumery, being used in some of the fine French creations.
2. Wood is used for posts, boards, veneers, furniture, decorative fittings, etc.
3. Fruits are eaten.
4. Flowers are tonic, stomachic and carminative and are used in dyspepsia, nausea and fever. They are also useful as a diuretic in renal diseases. The flower oil is used in cephalagia. Dried roots as well as the root bark are used as purgative and emmenagogue.

Black Dammar (*Canarium strictum* Roxb.)

1. The resin collected by hill tribes by the barbarous and destructive method of cutting in the lower part and then setting it on fire, is put to many uses, such as in the manufacture of blotting wax, varnishes, etc. It is also burnt as incense and is often used to expel mosquitoes out of dwelling houses. It is also used as a substitute for burgundy pitch in plasters.

Carissa carandas **L.**

1. Fruits are edible–suitable for tarts, salads, puddings
2. Useful for making dense hedges around fruit orchards as it forms an attractive theory hedge which can be trained to any height
3. The unripe fruit is sour and astringent and is used for pickles
4. Fruits are used to make refreshing juices or carbonated drinks
5. Ripe fruits are antiscorbatic, cooling and useful in bilious complaints–leaves are prescribed for remittent fever.

Michelia nilagirica **Zenk.**

1. Bark and leaves of the tree are considered febrifuge. The bark contains a volatile oil, acrid resins, tannin and a bitter principle. The flowers yield a volatile oil similar to the bark oil.

Sapindus emarginatus **Vahl.**

1. Fruits possess emetic, tonic, astringent and anthelmintic properties and are used in the treatment of asthma, colic due to indigestion, diarrhoea, cholera, tubercular glands, paralysis of the limbs and lumbago.
2. Fruit and root are used as fish poison.

Terminalia arjuna **(Roxb. ex DC.) Wight.**

1. Bark has been used for tanning and is suitable for heavy leathers.
2. The bark is arid and credited with styptic tonic, febrifugal and antidysenteric properties. In fractures and contusions, with excessive eechymosis, the powdered bark is taken with milk.
3. Powdered bark is used for relief in symptomatic complaint in hypertension.
4. Decoction of the bark is used as a wash in ulcers.

Indian Kino Tree (*Pterocarpus marsupium* Roxb.)

1. Bark is occasionally employed for dyeing
2. The exudate is collected and dried in the sun or shade and yield of dried gum is reported to be c.380 g. per tree
3. Bark is used as an astringent and flowers are used in toothache
4. Bruised leaves are useful as an external application for boils, sores and skin diseases
5. Leaves are excellent fodder–valuable as a manure in arecanut plantation

Balloon Vine or Blister Creeper (*Cardiospermum canescen* Wall.)

1. The tender, young shoots are used as vegetable and herb is used as fodder.
2. Used in rheumatism, lumbago, nervous diseases as a demulcent in orchitis and in dropsy.
3. Juice of the herb is used to cure earache and to reduce hardened tumours.
4. Used in preparing hair oil for curing dandruff, alopecia and for darkening hair.
5. The decoction of leaves is given for piles and as an infusion for general sores. The pungent vapour from the crushed leaves is inhaled to relieve headache. The juice is taken internally to reduce obesity.
6. Fresh leaf juice is given to asthma patients.

Lemon Grass [*Cymbopogon flexuosus* (Nees. ex. Steud.) Wats.]

1. Oil contains considerable amounts of resinous matter. Oil is used as perfumery, soaps and cosmetics and as a mosquito repellent
2. It is an important source of citral used in the production of a number of chemicals of industrial importance including ionones required for synthetic violet perfumes

***Piper mullesua* Buch.-Ham. ex D.Don.**

1. It is a substitute for *Piper betle*

2. As a masticatory, it is useful in catarrhal and pulmonary affections; it is also used for poultices.
3. Leaves contain a good amount of 'B' vitamins, ascorbic acid and carotene.

***Passiflora calcarata* Mark.**

1. Ornamentally used–grown as ornamentals in houses.

***Lepisanthes tetraphylla* (Vahl.) Radlk.**

1. The fruit has a sweetish pulp and is edible. The wood is white and moderately soft; it is occasionally used for house building.

***Styrax benzoin* Drayander.**

1. The resin, which possesses a pleasant balsamic odour, is obtained by making incisions in the stems of 6-7 years old wild or cultivated trees. A single tree yields c.10 kg. resin per year of its life. The resin is a pathological product which develops on incising the bark. The bark of a normal tree is rich in tanin and the resinotannols in benzoin are probably produced from the tanin. This is used entirely to perfumery and toilet preparations.
2. Benzoin is an irritating expectorant, carminative and diuretic, and forms an ingredient of inhalations in the treatment of catarrh of upper respiratory tract. It is also used to fix the odour of incenses, skin soaps, perfumes and other cosmetics and for fixing the taste of certain pharmaceutical preparations.
3. In veterinary practice, it is applied to indolent sores and foul ulcers.
4. It is also reported to be used in spirit–varnishes and in the manufacture of water proof laminated plastic sheets.
5. The extracts of bark and root of *S. benzoin* are said to possess insecticidal properties.
6. The wood is used for the construction of temporary buildings and is suitable also for match boxes.

***Elettaria cardamomum* (L.) Maton.**

1. Seeds are used as candiments and medicines.
2. Cardamom oil is used in several pharmaceutical preparations.

***Vernonia anthelmintica* (L.) Willd**

1. Seeds are used as an anthelmintic and to destroy insects.
2. The fruits are anti-inflammatory and used as tonic for asthma, leprosy, skin diseases, very specific for round worm and thread worm.

***Caryota urens* (Indian Sago Jaggery) L.**

1. The palm is cultivated for the starch to prepare sago, for toddy used in jaggery making and for fibre. It is also grown for ornament in gardens throughout India.
2. The major products obtained from the palm are sago, toddy, jaggery and fibre.
3. The trunk yields a flour from which a sago is made. The sago is reported to be equal in quality to the best sago of commerce obtained from *Metroxylon sagu* Rottb.
4. The palm is felled before flowering during the rainy season when it is dormant. The soft and starchy pith is separated, cut into pieces and pounded. These pieces are then mixed with water and strained through cloth. The settled starch matter is removed and dried. The yield varies between 100 and 150 kg per palm. The flour is a popular famine food in many parts of South India, particularly on the hills. The flour is valued for its cooling properties.
5. The bud or the cabbage is eaten raw or cooked with jaggery, or cooked as vegetable or even pickled.
6. The palm yields a large quantity of sap known as palm wine or sweet toddy from the time it begins to flower, when c.15 years old till it dies.
7. The sweet sap is largely used for making palmgur or jaggery. Other products obtained from the sweet toddy are palmolates, arrack, vinegar and food yeast.

8. A strong and valuable fibre is obtained from the fibre-vascular bundles found naked at the base of the leaf sheaths, petioles and floral stalks; the leaves also yield a fibre. On account of its strength and other properties, it is one of the best substitutes for bristle. It is used as a stiffener in admixture with bristles in brooms. The chief use of the fibre, extensively used in India, is in the manufacture of brushes, brooms, mattresses, baskets and bow-strings, hair, cloth, and horse brushes, brushes for polishing linen and cotton yarn and for cleaning scutched flax and brushing velvets.

Honey

1. Honey is collected by the tribals and sold. The peak season of collection is from March to May. They are sold in 750 ml. bottles.

The description of the plants is given in the following order; Botanical name with author citation, family, vernacular name, habit, parts used and how used.

Description of the Plants

The description of the plants is given in the following order:

1. Botanical name
2. Author citation
3. Family
4. Vernacular name (in Tamil Language)
5. Short description of the plant
6. Parts used
7. How used
8. Diseases cured or the importance of the plant

Abutilon indicum **(L.) Sweet ssp. *indicum* [Malvaceae]**

Thuthi chedi or Norupan; erect velvety tomentose undershrub; young shoots and leaves are used; four young shoots of this plant are munched in the mouth with seven garlic cloves and the air is blown into the ear of the affected person for abdominal colic and gastritis; stream crabs are boiled and ground with these leaves and everything is put in a white cloth and that heat is applied on the body for swelling and body pain.

Acacia caesia (L.) Willd. **[Mimosaceae]**

Indu kodi; shrub. A climbing shrub with (large prickles), dense foliage of Phyllodia and yellowish flower in cylindrical spikes, NTFP.

Acacia canescens Grah. **[Mimosaceae]**

Indang kirai; climbing shrub; leaves and bark are used; leaves are cooked and eaten as greens for general body health; the leaves are dried and sold at Rs.3/kg; bark is powdered and sprayed in the river or stream to catch fishes.

Acacia pennata (L.) Willd. **[Mimosaceae]**

Indan kirai; straggling shrub with bipinnate leaves; leaves are cooked and eaten as greens for general body health and especially for digestive disorders.

Acacia sinuata (Lour.) Merr. (*A. concinna* (Willd.) DC.) **[Mimosaceae]**

Sigakai (Soap nut tree); prickly scandent shrubs with bipinnate leaves; NTFP.

Acalypha fruticosa Forssk. **[Euphorbiaceae]**

Chinni chedi; glandular aromatic shrub with short axillary spikes; leaves are ground and applied on the affected place for skin problems.

Acalypha indica L. **[Euphorbiaceae]**

Kuppai meni; erect herb with long petioled rhomboid leaves; leaves are ground, applied on the wounds, dried and taken bath later in warm water for scabies and dermatitis; steam is inhaled with these leaves for asthma, cold and cough.

Acalypha racemosa Heyne ex Baill. **[Euphorbiaceae]**

Chinni thalai; shrub; leaves are used; mixed with the leaves of *Acthyranthes aspera*, gingeli oil, and small onion, ground, fried and applied on the skin of small children for itching on the body and dermatitis.

Achyranthes aspera L. var. *aspera* **[Amaranthaceae]**

Nayuringi; erect undershrub with ovate leaves; whole plant and leaves are used; decoction is made with the whole plant and given inside for renal dropsies, diarrhoea, anaemia, piles, abdominal disorders; leaves are crushed, the extract is put in the palm of the delivering mother and licked for safe delivery.

***Acorus calamus* L. [Araceae]**

Vasambu; aromatic marshy herb with stout rhizome and distichous leaves; rhizome and bark are used; rhizomes are ground and given in for mental ailments, chronic diarrhoea, dysentery, abdominal tumours; bark is mixed with cumin seeds, pepper, *Alpinia calcarata*–all ground together–pepper water is made and drunk–good for delivered mothers.

***Adhatoda zeylanica* Medic. (*A. vasica* (L.) Nees) [Acanthaceae]**

Adathoda; ervergreen branched shrubs; Leaves are used; Leaves are dried in shadow, ground, added with honey and taken inside for cold, cough, throat infection; these leaves are mixed with the leaves of *Vitex trifolia* and boiled in water and taken bath to remove tiredness of the body; some leaves are made warm, put as pillow and slept on it to reduce heaviness of the head; fresh or dried leaves are powdered and taken in to cure bronchial troubles; juice of the leaves is taken orally for diarrhoea, dysentery, glandular tumours; powdered leaves are used externally for skin affections.

***Adiantum raddianum* Presl. [Adiantaceae]**

Pachai paruva thalai; herb; leaves are squeezed and the extract is applied at the place of warts and pimples.

***Aerva lanata* (L.) Juss. [Amaranthaceae]**

Poolam poova chedi sirupalai; erect herb or undershrub; flower and leaves are used; this flower is tied in cowsheds, houses, temples as a sign of village feast and tied in cultivation fields to be very auspicious; Leaf extract is taken and given for cough, sore throat, diabetes.

***Agaricus campestris* [Agaricaceae (fungi)]**

Kalan; saprophytic, hyphal, much branched; cooked and eaten.

***Agave cantala* Roxb. (*A. americana* Auct. non. L.) [Agavaceae]**

Karungathalai; bushy herb without stem; the pith in the stem or the core of the plant is mixed with castor oil and eaten daily for any stomach disorder, to prevent the effect of black magic and evil eyes, veneral sores, syphilis and dysentery.

Ageratum conyzoides L. [Asteraceae]

Vettu kayathalai or Kaliamman thalai; shrub; Leaves are ground and applied on the head before taking bath for head ache; leaves are ground, mixed with the sap of banana tree and lime powder, when the mixture becomes yellowish is applied on the wound and dressed for cut injuries and other wounds; the leaves are ground with salt, small onion and a small piece of washing soap and applied at the affected place for wounds, cut injuries, boils in hands and legs, paranchia.

Ailanthus excelsa Roxb. [Simaroubaceae]

Pinari (Maharukh); large tree with light grey bark; collected and used as firewood.

Albizia amara (Roxb.) Boiv. [Mimosaceae]

Unjal arapu maram; medium-sized tree with yellow, fragrant flowers in heads; leaves are ground, applied on the head and taken bath to reduce body heat.

Albizia chinensis (Osbeck) Merr. [Mimosaceae]

Arappu maram; large deciduous tree; leaves are ground, applied on the head, allowed to dry for some time and taken bath to reduce the falling of hair and to blacken the hair.

Aleurites moluccana (L.) Willd. [Euphorbiaceae]

Kattu kottai chedi or Amanakku; evergreen tree with long petioled leaves; sap and bark are used; the sap is taken inside; mouth is gargled with the sap for stomach and mouth ulcers; the teeth is brushed with the bark for mouth ulcer or angular stomatitis.

Allium cepa L. [Alliaceae]

Chinna vengayam (small onion); bulbous herbs with sub-distichous leaves; Root bulb is used in combination with other ingredients to cure commonly occuring sicknesses.

Allium sativum L. [Alliaceae]

Vellai poondu (Garlic); perennial bulbous herb with flat leaves; root yam; garlic is crushed and eaten or boiled and eaten for tuberculosis, cardiac diseases, constipation, indigestion, abdominal disorders, worm infestation; garlic is boiled in gingelly oil and poured into the ear for ear pain.

Aloe vera **(L.) Burm. f. (*A. barbadensis* Mill.) [Liliaceae]**

Chothu kathalai; perennial herb with fleshy ensiform leaves; leaves and pith are used; the young shoots are ground with salt and applied externally at the affected place for joint pain and swelling in the joints; the pith is fried in coconut oil and eaten for dry cough.

Alphonsea sclerocarpa **Thwaites [Annonaceae]**

Thutta maram; tree; fruits are collected and eaten as food.

Alpinia calcarata **Roscoe [Zingiberaceae]**

Chitharathai; erect rhizomatous herb with distichous linear leaves; this rhizome, in combination with gallnut, *Acorus calamus*, aricanut, *Ferula asafoetida*–all ground, mixed with mother's milk and given in for good digestion, stomach disorders in children.

Alternanthera sessilis **(L.) R. Br. ex DC. [Amaranthaceae]**

Ponnaankanni; prostrate herb with fleshy leaves and white flowers; leaves; leaves are cooked as greens and eaten for general health of the body especially for fever and haemorrhage, to cool down the body and for good eye sight; the leaves are ground with camphour and appplied on the forehead at night and given warm water bath on the following morning for migrane.

Alternanthera pungens **Kunth. [Amaranthaceae]**

Ottara mul; creeping herb (weed); leaves are boiled and eaten as greens for general body health.

Alternanthera viridis **L. [Amaranthaceae]**

Moochu kuthu thalai or kuppai keerai; erect herb, glabrous; leaves ovate or deltoid; flowers in axillary and terminal; leaves; these leaves are squeezed with one onion in the palm and that is kept in the palm and air is blown for gastritis and muscle catches in the body.

Alysicarpus vaginalis **(L.) DC. var. *vaginalis* [Fabaceae]**

Rama kirai; robust much branched suberect herb with glabrous leaflet; leaves are cooked and eaten as greens for general body health.

Amaranthus caudatus **L. [Amaranthaceae]**

Sirukeerai; erect herb; cooked and eaten as greens.

***Amaranthus spinosus* L. [Amaranthaceae]**

Mullukkirai; erect herb with hard spines; root, leaf and whole plant are used; they are cooked and eaten for burning sensation, snake-bite poisoning, amenorrhoea, leucorrhoea, anuresis, abscess.

***Amaranthus tricolor* L. (*A. mangostanus* L.; *A. gangeticus* L.) [Amaranthaceae]**

Tandukkeerai; erect stout herb with leaves of various colours; cooked and eaten as greens.

***Amaranthus viridis* L. [Amaranthaceae]**

Siru keerai; herb; leaves are cooked and eaten as greens to add essential vitamins to the body.

***Ananas comosus* (L.) Merr. (*A. sativus* Schult. & Schult. f.) [Bromeliaceae]**

Annaci (Pine apple); perennial herbs with stout stems; the juice from the fruit is mixed with water and taken to kill the worms in the stomach and to cure indigestion.

***Anaphalis lawii* (Hook.f.) Gamble [Asteraceae]**

Thee chutta kaya chedi; tall herb; the leaves are ground and applied externally on the burn wounds.

***Anisochilus carnosus* (L.f.) Wall. ex Benth. [Lamiaceae]**

Nai semarai; annual erect herbs with 4-angled stem; leaves; leaves are slighlty burnt in the fire and rubbed against sores in the leg for intertrigo; leaves are ground and applied on the body of dogs and given bath for rabies; ground leaves are applied on the body of affected persons for dermatitis and scabies.

***Anogeissus latifolia* (Roxb. ex DC.) Wall. [Combretaceae]**

Vekkoli maram; deciduous tree with smooth white bark; leaves are ground with small onion and applied at the place of impetigo.

***Aristida setacea* Retz. [Poaceae]**

Velakkamathu seeval chedi; herbs; entire plant; the entire plant is collected, tied together and used for sweeping.

***Aristolochia tagala* Cham., B.C. Stone & K.N. Nair [Aristolochiaceae]**

Kiradam pattai; large climbing shrubs with grooved stem; bark and root are used; the bark is ground with water and given inside for stomach ache; the root is ground and given inside for extreme stomach pain.

***Artemisia nilagirica* (C. B. Clarke) Pamp. (*A. vulgaris* auct.non L.) Asteraceae]**

Pachchaippu chedi; tall aromatic herb; the leaves are squeezed in the palm and the extract is taken inside for ordinary cough and cold.

***Artocarpus heterophyllus* Lam. (*A. integrifolius* auct. non L.f., 1781) [Moraceae]**

Pala maram (Jack tree); Laticiferous large tree used as firewood.

***Arundo donax* L. [Poaceae]**

Nanal; Tall stout herb with creeping rhizome and large panicles; leaves and roots are used; Leaves are used to make thatched sheds; Root is ground and applied externally for wound, skin diseases, burning sensation.

***Asparagus racemosus* Willd. [Asparagaceae]**

Neevakka mullu; Perennial tuberous scandent shrub with triquetrous cladodes and stem with spines; Sap/latex, leaves and yam are used; The sap is taken from the young shoot and applied at the place of stye; Leaves slightly heated in fire and tied at the place of paranychia; Yam is cooked and eaten to prevent kidney stone formation and burning micturtion.

***Asystasia gangetica* (L.) T. Anders. (*A. coromandeliana* Nees) [Acanthaceae]**

Valukka kirai; Erect, perennial herb with four-angled stem; leaves are cooked and eaten as greens for general body health.

***Azadirachta indica* A. Juss. [Meliaceae]**

Vembu (Neem); Large tree with leaves crowded at the ends of branchlets; bark, oil, leaves, and fresh tender twigs are used; bark is ground and applied externally for skin troubles; oil is extracted and applied externally on the skin for skin troubles and given to drink for the delivered mothers to clean the uterus and stomach; The leaves are ground with the leaves of *Cyanodon dactylon* and turmeric and applied externally for mumps, chicken pox; this poultice can also be given inside; the decoction is also made and drunk; the decoction is given for ulcers and eczema.

Baeolepis nervosa **(Wight & Arn.) Decne ex Moq. in DC. [Periplocaceae]**

Mavaali kilangu; climber; root; the root is ground with tamarine, pepper and salt, fried in cooking oil and taken in every day for general body health. This preparation can be kept for a long-time.

Bambusa bambos **(L.) Voss (*Bambusa arundinacea* (Retz.) Willd.) Roxb. [Poaceae]**

Moongil (Bamboo); Arborescent thorny grass with hallow shining culms; bamboo rice and pith are used; bamboo rice is cooked and eaten-nutrition for the body; the pith is taken, sharpened, soaked in rice boiled water, that is kept outside in dew for one right, this is eaten by women early in the morning in empty stomach to produce sterility in women–(to avoid conception of child).

Barbula convoluta **[Bryophyta]**

Tree moss; Gametophytic plants, consists of rhizoides, stem and leaves, peristome spirally twisted; NTFP.

Basella alba **L. var. *alba* [Basellaceae]**

Kodi pasali; twining herb with broadly ovate leaves; the leaves are washed, fried in cooking oil and eaten. The leaves can be ground and eaten to reduce body heat, to cure scabies for children, to cure stomach disorders and problems and constipation.

Begonia floccifera **Bedd. [Begoniaceae]**

Kalthamarai; herb; flower used; The flower is ground and added in cow's milk and given to cure VD, STD and leukorrhoea (too much of white discharge for woman).

Begonia malabarica **Lam. [Begoniaceae]**

Parai pattai; shrub, much branched, succulent; the leaves are fried and eaten to reduce heart pain, chest pain and to increase blood circulation.

Benincasa hispida **(Thunb.) Cogn. (*Cucurbita hispida* (Thunb.) Cogn.; *Benincasa cerifera* Savi.) [Cucurbitaceae]**

Sambal pusani; climbing shrub with hirsute branchlets; commonly used vegetable for food.

***Bentinkia condapanna* A. Berry [Arecaceae]**

Koonthapanai; tree and yam are used; tree and yam are cooked and eaten for general body health.

***Beta vulgaris* L. [Chenopodiaceae]**

Beet root; climbing herb; commonly used vegetable for food.

***Bidens pilosa* L. [Asteraceae]**

Ottu chedi; herb; the leaves are squeezed in the palm and the juice is taken inside for menorrhagia; These leaves with the leaves of *Phyllanthus amarus, Solanum laeve, Diplocyclos palmatus* are boiled and eaten for tiredness and weakness felt in the body.

***Blepharis maderaspatensis* (L.) Heyne ex Roth (*B. boerhaviifolia* Pers.) [Acanthaceae]**

Thonimalai marundhu; Prostrate hairy herb; entire plant especially the leaves are used; entire plant is ground and tied at the affected place after setting it right for fracture or dislocation; The leaves are ground and given inside for fracture or dislocation; The leaves of this plant are mixed with the leaves of bamboo, that is mixed with country hen's egg and taken inside for blunt injury.

***Blumea mollis* (D. Don) Merr. [Asteraceae]**

Kaliamman thalai; Herb; leaves are used; leaves are squeezed in the palm and the extract is drunk regularly to avoid pregnancy; Leaves are ground and applied externally for curing cut injury and other wounds; The extract is given after delivery to bring out the retained placenta and also clean the uterus after delivery.

***Blumea oxyodonta* DC. [Asteraceae]**

Thee chutta kaya marundhu; prostrate herb with radiating branch; leaves are used; Leaves are ground and applied on the burn wounds for children to cure burn wounds.

***Blumea virens* Wall. ex DC. [Asteraceae]**

Pasan pachai ilai; herb with pilose stem and branch; leaves are used; leaves are ground and tied on the impetigo. It wiil make it ripen and bring out the dirt and puss; the juice is squeezed out and given to delivered mothers for the placenta to come out easily.

Boerhavia diffusa **L. [Nyctaginaceae]**

Thanneer kodappan kirai; Diffuse herb; leaves and root, seed, whole palnt are used; The leaves are boiled and eaten as food. The leaves are spread on the floor and babies/children are made to sleep on that for general body health and to reduce body heat. The root or seed or whole plant is boiled and eaten for edema, urinary disorders, anemia, cardiac disorders and stomach pain.

Brassica juncea **(L.) Czernj. and Coss. [Brassicaceae]**

Kadugu keerai; Erect herb with lyrate leaves; root and leaves are used; The root is boiled with small onion and cumin seeds and that extract is drunk for abdominal pain after delivery and any urinary infection and urinary retention; leaves are squeezed with salt in the palm and that extract is poured into the ear for pain in the ear and otorrhoea (C.S.O.M.)

Brassica nigra **(Koch) Deutsch. [Brassicaceae)**

Kadugu kirai; erect herb; cooked and eaten as greens.

Brassica oleracea **L. var.** ***capitata*** **(L.) Alef. [Brassicaceae]**

Muttai kosu (Cabbage); herb with leaves forming the head; commonly used vegetable for food.

Brassica oleracea **var.** ***botrytis*** **L. [Brassicaceae]**

Cauli flower; herb; commonly used vegetable for food.

Buddleja asiatica **Lour. [Buddlejaceae]**

Paruppu kirai; large shrub; leaves are cooked and eaten for general body health.

Cajanus cajan **(L.) Millsp. (*C. indicus* L.) [Fabaceae]**

Tauvarai; shrub with trifoliate leaves. The young shoots are grown in mud pots and offered to the diety during village annual festival.

Calamus rotang **L. [Arecaceae]**

Vasambu; tall slender spiny climbing shrub with truncate, short ochreae; fresh root is used; fresh root is ground, mixed with water and eaten for stomach disorders and ache and rheumatism.

***Calotropis procera* (Ait.) R. Br. [Asclepiadaceae]**

Erukkalam; erect shrub with soft, corky, spongy bark; sap is used; the sap is applied on the nail gaps in the fingers and toes to reduce body heat and hot flesh.

***Canarium strictum* Roxb. [Burseraceae]**

Karuppu Kungiliam (Black Dammar); large tree; branches and leaves beneath rufous-tomentose; NTFP.

***Cansjera rheedii* Gmel. [Opiliaceae]**

Palivi kirai; climbing shrub; leaves are used; leaves are cooked and eaten as greens for general body health; leaf extract is given to delivered mothers to clean the uterus after delivery.

***Canthium dicoccum* (Gaertn.) Tejism. & Binn. var. *dicoccum* [Rubiaceae]**

Sulandu maram or Vippuruthi maram; unarmed tree; stem, leaves and bark are used; the fresh bark is peeled off and the stem is burnt and used as light to walk in the forest at night; the leaves are boiled and that heat is applied on the affected part for abdominal problem, pulging, swelling; the bark is dried, ground and applied externally for fractures.

***Capsicum frutescens* L. [Solanaceae]**

Kattu milagai chedi; annual herb; leaves, fruit and seed are used; leaves are cooked and eaten for general body health; leaves are ground with lime powder and applied on the forehead for head ache; fruits are fried in oil, ground and applied externally for wounds and cut injuries.

***Cardiospermum canescens* Wall. [Sapindaceae]**

Kochan kodi/Mudakathan; slender tendril climber with pubescent leaves; decoction is made with the entire plant and given to the affected person to cure rheumatism; the entire plant is cut into small pieces, dried and sold at Rs.10/kg; fresh plants are sold at Rs.5/kg- this plant is available at all seasons; during cattle pongal, this plant is tied in the cattle shed, and tied around the neck of the cattle; this plant is also used for 'Jallikkattu' (a special celebration with cattle).

***Cardiospermum halicacabum* L. [Sapindaceae]**

Muduku thalai or Mudukkuthan thalai; Tendril bearing climbing herb; leaves are used; leaves are boiled with other condiments and that boiled water is distilled and drunk for ordinary fever, edema, myalgia, tiredness of the body; the leaves are dried in shade and powdered and one spoon full of that powder is taken inside every morning in empty stomach; pepper water can also be prepared with this powder and drunk to relieve body pain.

***Carica papaya* L. [Caricaceae]**

Pappali (Papaya); tree with pinnatifid leaves; sap and fruit are used; sap is applied in the affected place for swelling in the gum; raw fruit is ground with mustard seed, neem leaves, palmyra jaggery and mixed with gingeli oil and given for two days to induce abortion.

***Carissa carandas* L. [Apocynaceae]**

Kila maram; scandent shrub with straight or slightly curved spines; leaves and flowers are used; ripe fruits are eaten and raw fruits are used to prepare pickle for food; decoction of the leaves is given inside for intermittent fever; the flowers are boiled and given inside for indigerstion and hypertension; fruits are collected and sold as NTFP.

***Caryota urens* L. [Arecaceae]**

Koontha panai; lofty tree with smooth trunk; leaves and inflorescence and root pith are used; the leaves are used to decorate the pandals for weddings and other important functions and celebrations. Sweet toddy is tapped from young inflorescence if the tree is more than 15 years old. Jaggery is made out of this. Root pith is eaten; The leaves are used to put up thatched sheds. The inflorescence is also used as decorative material.

***Cassia leschenaultiana* DC. [Caesalpiniaceae]**

Keela nalli; bushy herb; leaves are used; the leaves are crushed and squeezed and the extract is taken inside. The leaves are also boiled with the leaves of *Phyllanthus amarus*, *Solanum laeve*, *Diplocyclos palmatus* and eaten for body pain.

***Cassia sophera* L. [Caesalpiniaceae]**

Ponnavarai; diffuse foetid undershrubs; leaves are used; the leaf is shown in fire a little after applying castor oil on it and stuck at the place of impetigo.

***Casuarina litorea* L. (*C. equisetifolia* L.) [Casuarinaceae]**

Chavuku; tall tree with cylindric or angled branchlets; the latex is collected and applied to wounds of any type.

***Centella asiatica* (L.) Urban [Apiaceae]**

Vellarai; trailing herb rooting at nodes; leaves are used; a handful of leaves is squeezed with salt, eight drops are put in the palm and eight drops in the upper palm to be licked-this is to be done three times on consecutive days on the third or 11th day after mensus. It helps women to conceive. The leaves are boiled in water and the water is distilled and drunk to cure any stomach disorder, worm infestation and this extract is given to pregnant women for strength and safe delivery; the leaves are boiled in water and kept soaked for one night and that water is distilled and drunk on the following morning in empty stomach for stomach ulcer and stomach disorders.

***Chamaecrista absus* (L.) Irwin & Barnerby (*Cassia absus* L.) [Caesalpiniaceae]**

Avaran keerai; glandular hairy herb; flower, root and seeds are used; the flowers are boiled with *Hibiscus* flower and banana flower and the decoction is distilled and taken inside for menorhagia, hernia, indigestion and bloated abdomen; the root is burnt, powdered, mixed with honey and given to small children for indigestion and other stomach problems; seeds are ground and applied externally fo skin troubles.

***Chenopodium ambrosoides* L. [Chenopodiaceae]**

Natha chedi or Mannanai chedi; aromatic, erect, much branched herb; leaves are used; the leaf is slightly burnt in fire, gingeli oil is applied to it and tied on the wound for any type of wound.

***Chrysanthemum indicum* L. [Asteraceae]**

Pachai poo chedi; erect herb with yellow flowers; leaves are used; leaves are ground and given inside to reduce body heat and migrane.

***Cicer arietinum* L. [Fabaceae]**

Pottukkadalai; erect glandular-hairy herb with pinnate leaves; the seeds are ground with mustard seed and leaves of *Trachyspermum ammi* and taken inside to induce abortion.

Cinnamomum macrocarpum **Hook. f. [Lauraceae]**

Lavanga maram; medium-sized tree; bark is added to curry in a small amount for abdominal colic and gastritis; bark is ground, applied from head to foot, dried and taken bath for body pain; this bark and ginger are added to coffee and drunk to relieve body pain.

Cinnamomum zeylanicum **Blume. [Lauraceae]**

Karimasal pattai; tree; bark is used; the bark is added with other ingredients and the extract is taken in for fever, headache, head being heavy. It is good for delivered mothers.

Cipadessa baccifera **(Roth) Miq. [Meliaceae]**

Semitti chedi; small tree; leaves are squeezed in the palm and the extract taken inside for insect bite and other poison in the body.

Cissampelos pariera **L. var. *hirsuta* (Buch.-Ham. ex DC.) Forman [Menispermaceae]**

Urikkakodi or Suruli ver; villous twining herb; root and leaves are used; root is dried, powdered and given to small children for indigestion, bloated abdomen, intestinal obstruction; the root is squeezed, extract is taken out and drunk for any stomach disorder and poison in the body; the smell of the root drives away snakes to guard from poison bite; the leaves are ground and given inside for continuous stomach ache.

Cissus quadrangularis **L. [Vitaceae]**

Perandai; Scandent shrub with quadrangular stem; stem and leaves are used; boiled and added with curry and eaten for osteo arthritis; the juice is squeezed out of the stem and taken in to cure stomach problems, stomach upset, indigestion, lack of apitite. This also can be given to cattle for the same problems. The leaves are fried and eaten for arthritis and joint pain.

Citrus aurantium **L. [Rutaceae]**

Aaranju (orange); trees with greenish-white stem. The fruits are eaten.

Citrus limon **(L.) Burm. f. (*C. medica* L. var. *limonum* (Risso) Hk. f.) [Rutaceae]**

Malai Elumichai; small tree; fruits are used; ripe fruit is inserted into the affected finger for paranchia; the juice is mixed with salt and given. The juice is mixed with fresh cow's milk or goat's milk and given in for stomach pain and dysentery; the juice is mixed with honey and taken in warm water to cure cough.

Citrus pennivesiculata (Lush.) Tanaka [Rutaceae]

Siru kolingi kai; tree; fruit and leaves are used; the fruits are squeezed with pepper and the extract is taken in by the patient by sitting on sand for cold and running nose; leaves are eaten raw or boiled for general good health of the body.

Clausena heptaphylla (Roxb.) Wight & Arn. [Rutaceae]

Pottichedi; small tree; entire plant is used; the sticks from this plant are used to put up bathing sheds; these sticks are used to put up thatched sheds to keep the girls who have attained puberty, in seclusion for a few days. Significance of this plant: this plant is closely associated with their culture; this plant is considered auspicious because it gives out sap when cut.

Clausena indica (Dalz.) Oliver in J. [Rutaceae]

Theva potti maram; small tree; branches with leaves; this plant is used for worship in temples; as a sign of chicken pox infection, this plant is placed in front of the house.

Clematis munroiana Wight. [Ranunculaceae]

Chaya kodi or Pomplikuodi; climbing shrub; young shoots and fresh leaves are used; the young shoots and fresh leaves are squeezed and the smell inhaled through the nostrils for running nose, cough and cold; the leaves are squeezed in the palm and the extract is applied on the affected part for skin diseases.

Cleome gynandra L. (_Gynandropsis pentaphylla_ DC.) [Cleomaceae]

Kattu kadugu; erect branched hairy herb; seeds and leaves are used; seeds are used to make curry; leaf juice is applied externally for headache, neuralgia, rheumatism and other local pains.

Cleome monophylla L. [Cleomaceae]

Malai kadugu; herb; seeds and leaves are used; the seeds are cooked with small onion and eaten for dyspnoea; leaves are boiled with dal, ground and eaten for general body health.

Clerodendrum serratum **(L.) Moon. (Verbenaceae)**

Malai errukkam; large shrub with quadrangular stems; root and leaves are used; root is powdered and given inside by mixing with water; the powdered root can also be kept for a long time for fissure-in-ano, piles; the leaves are ground and applied all over the body before sleeping for anaemia, myalgia.

Coccinia grandis **(L.) Voigt. (*C. indica* Wt. & Arn.) [Cucurbitaceae]**

Kovan kirai; scandent or prostrate much branched tuberous shrub; leaves are boiled and eaten as greens for general body health.

Cocos nucifera **L. [Arecaceae]**

Thennai maram (Coconut tree); tall tree with annulate trunk; fruit and oil are used; a piece of coconut is dipped in neem oil, heated in lantern fire and applied on the affected tooth for tooth ache; one tea spoon of oil is mixed with one mug of water, salt is mixed, that water is warmed-when the water is sufficiently warm, mouth in gargled for tooth ache and decayed tooth.

Coffea arabica **L. [Rubiaceae]**

Kappi chedi; shrub or small tree with cariaceous opposite leaves; collected and used as firewood. (The seeds are used as beverages).

Coffea robusta **L. (Rubiaceae)**

Kappi chedi; large shrub or small tree; collected and used as firewood. (The seeds are powdered and used to prepare beverages).

Colocasia esculenta **(L.) Schott (*C. antiquorum* Schott) [Araceae]**

Semban kirai; rhizomatous herb; leaves are boiled and eaten as greens to remove kidney stone and prevent the formation of kidney stone.

Commelina benghalensis **L. [Commelinaceae]**

Thengai pottu kirai or amala ilai; procumbent herb with cleistogamous fertile white flowers; the leaves are boiled, cooked and eaten as greens; the leaves are also fried in combination with cumin seeds, gingeli oil, small onion, salt, dry chilly and eaten; this food adds essential vitamins to the body; this also cures piles.

Commelina hirsuta **(Wight.) C.B. Clarke. [Commelinaceae]**

Valukka keerai; branched herb; leaves are cooked and eaten as greens for general body health; leaves are ground with black gram, fried in gingeli oil and swallowed for chest pain and wheezing; leaves are squeezed and the extract taken inside for stomach disorders and diarrhoea.

Commiphora caudata **(Wight & Arn.) Engl. [Burseraceae]**

Kiluvai maram; tree; the wood is used to make pounder and pounder base.

Corchorus aestuans **L. (*C. acutangulus* Lam.) [Tiliaceae]**

Chinni chedi; suffrutescent herb; entire plant; a little portion of the plant is plucked and kept at the centre of the upper head to prevent evil thoughts of others from affecting us; If unknowingly, one stamples on things used for black magic, it will not have any effect.

Cordia obliqua **Willd. var. *tomentosa* (Wall.) Kazmi [Boraginaceae]**

Thalingi maram; small deciduous tree; leaves, mixed with lime powder are ground and applied at all the paining joints for joint pain and body pain.

Coriandrum sativum **L. [Apiaceae]**

Kothamalli; herb with lower dissected and upper entire leaves; young leaves and seeds are used; the leaves are ground with salt and eaten for gastritis and piles; the seeds are used to prepare pepper water and it is good for general health.

Crassocephalum crepidioides **(Benth.) S. Moore. [Asteraceae]**

Kannaruppan thalai; annual herb; stem or leaves; the leaves are ground and taken inside for too much of loose motion and dysentery.

Crotalaria pallida **Dryand [Fabaceae]**

Chandai pidichi kai; under shrub; entire plant; this plant is taken and tied in front of the house of enemies; they will automatically pack their things and desert their house.

Croton bonplandianus **Baill [Euphorbiaceae]**

Rail poondu; erect herb; the latex is applied at the place of impetigo.

Cryptolepis buchanani **R. & S. [Asclepiadaceae]**

Paal kodi or marappalai chedi; twining herb; bark or the entire plant; when going to ask for a bride, if the bride's family declines-then this plant is taken while going for the second time, it is believed that the bride's family would accept.

Ctenolepis garcinii **(Burm. f.) C.B. Clarke. [Cucurbitaceae]**

Rama kirai; pretty climber; leaves are cooked and eaten to remove tiredness of the body.

Cuminum cyminum **L. [Apiaceae]**

Siragam (Cumin seeds); slender herb with pinnately dissected leaves; seeds; the seeds are ground and mixed with water and taken inside for dysentery; the seeds are fried, soaked in warm water and eaten. It is good for pregnant mothers.

Curcuma longa **L. [Zingiberaceae]**

Manjal-turmeric; rhizomatous herb with sessile cylindric tubers; rhizome and powdered rhizome; rhizome is mixed with neem leaves, leaves of *Cynodon dactylon*–all ground and applied externally and also given inside for sprain, bruises, mumps and chicken pox; rhizome is burnt in fire, covered in a cloth and the smoke is inhaled through the nostrils for migrane head ache; rhizome is tied in a white cloth, burnt in fire, ground, mixed with water and given inside for uncontrolled loose motion; one water frog is cut open in the middle, a little turmeric powder is put inside, the head and the legs are removed from the frog and that is tightly tied at the place of impetigo.

Cynoglossum zeylanicum **(Vahl. ex Honem.) Thunb. ex Lehm. [Boraginaceae]**

Ottukka chedi; erect herb; leaves; leaves are ground and given inside; this plant can also be tied around the neck of the affected person to cure insanity and other mental illness.

Cyathula prostrata **(L.) Bl. [Amaranthaceae]**

Parappu kirai; procumbent herb; leaves and root; the leaves are boiled with yellow gram (dal), ground and eaten for general body health; decoction is made using the root for dysentery and skin complaints.

Cymbopogon flexuosus (Steud.) P. Watson [Poaceae]

Thaila pullu; herb; leaves; the leaves are boiled with ginger and pepper and that decoction is drunk; if a hundred kilos of lemon grass is boiled, 750 ml. of oil can be extracted; this oil can be applied on the body for head ache and fever; if oil bath is taken with this oil, it relieves body pain and tiredness; for stomach pain, two drops of this oil are added to a glass of water and drunk.

Cymbopogon martinii (Roxb.) Wats [Poaceae]

Kanam pullu; tufted aromatic grass; used to thatch houses.

Cymbopogon nardus (L.) Rendle. var. _confertiflorus_ (Stend.) Stepf ex Bor. [Poaceae]

Kaavattum pollu; a tall aromatic herb; used to thatch houses.

Cymbopogon travancorensis Bor. [Poaceae]

Tharuvai pullu; an aromatic perennial herb; used to thatch houses.

Cynodon dactylon (L.) Pers. [Poaceae]

Arugampul; creeping herb with wiry slender calms; leaves (rhizome and the whole plant); rhizome is powdered and given in for fenito-urinary troubles; a handful of this grass is ground, distilled and that extract is taken early morning in empty stomach for haemorrhage, fainting, burning sensation, menorrhagia, distaste, vomiting, to prevent heart attacks and heart problems, purify blood, increase the production of new blood cells; this grass is ground with the leaves of neem, turmeric and applied on the face for mumps; this grass is mixed with the young shoots of *Arunda donax* and one crystal salt-all together ground and applied all over the body and taken bath for itching, heat boils, scabies, boils and scratches around genitals.

Cyphomandra betacea Sendt. [Solanaceae]

Marathakkali; small tree, fruits edible; leaves and fruits are used; the leaves are cooked and eaten; the skin of the fruit is removed and eaten raw for stomach and mouth ulcers.

Datura arborea L. [Solanaceae]

Umathai; large shrub; flower; neem oil is applied in a pure white cloth, seed-removed dry chilly is rolled in the cloth, the entire thing is put into this flower-fire is put at the tip of the flower; the smoke is inhaled through the nostrils-done for three days for migrane head ache.

***Datura innoxia* Mill. [Solanaceae]**

Umathai; glandular tomentose herb or undershrub; leaves; the leaves are ground, applied all over the body and taken bath in warm water for body pain and to remove tiredness and also to reduce the effect of witchcraft and black magic.

***Datura metel* L. (*D. fastuosa* L. var. *alba* (Nees) Clarke) [Solanaceae]**

Karu umathai; under shrub; leaves and flowers; the leaves are ground and applied all over the body before taking bath; the flowers are ground and applied all over the body before taking bath to remove tiredness of the body and myalgia.

***Desmodium triflorum* (L.) DC (Fabaceae)**

Sakkarai kodi; wiry spreading herb rooting at nodes; leaves; these leaves are ground with small onion and given inside and also applied externally for general weakness of the body and to control body becoming weak due to diseases, dysentery and diarrhoea.

***Daucus carota* L. [Apiaceae]**

Karatu (Carrot); hairy tuberous herb with pinnatisect leaves.

***Didymocarpus innominatus* Burtt. [Gesneriaceae]**

Kal Ilai; a large scapigerous herb with many scapes; leaves; the leaves are squeezed and the extract is applied all over the body thrice for anemia, fainting attack and giddiness; the leaves are ground with small onion, applied on the head. Head is covered with a white cloth for some time and then given bath; this is done three times to remove dandruff from the hair.

***Dioscorea alata* L. [Dioscoreaceae]**

Mullu valli kilangu; tuberous climber with quandrangular winged stem; cooked and eaten as greens.

***Dioscorea bulbifera* L. var. *bulbifera* [Dioscoreaceae]**

Vallikilangu; climber with stem twining to left bearing bulbils in leaf axils; cooked and eaten.

***Dioscorea esculenta* (Lour.) Burkill [Dioscoreaceae]**

Siruvalli kilangu; tuberous prickly climber, stems twining to left; tuber; the yam is dug out, boiled and eaten; good for general body health and to cure piles.

***Dioscorea pentaphylla* L. var. *pentaphylla* [Dioscoreaceae]**

Mullu kilangu; climber with stems twining to left; cooked and eaten.

***Diospyros nigricans* Dabiz. [Ebenaceae]**

Vakkanai; evergreen tree; bark; the bark is powdered, soaked in water and the face is washed with this water for night blindness; the bark is mixed with sand, ground and sprayed to catch fishes; the fishes will faint and float in the water.

***Diplocyclos palmatus* (L.) C. Jeffrey. (*Bryonopsis laciniosa* sensu Naud.) [Cucurbitaceae]**

Aiviral keerai; tuberous climbing herb with forked tendrils; leaves; leaves are cooked and eaten as greens; this adds essential vitamins to the body: this helps to remove tiredness of the body, relieves body pain especially in the hands and legs and used for myalgia.

***Dodonaea viscosa* Jacq. [Sapindaceae]**

Velari chedi; bushy shrub or small tree; leaves; leaves are ground and applied at the affected place for burns, swellings, wounds, head ache.

***Dumasia villosa* DC. [Fabaceae]**

Katta kila kodi; climber; leaves and bark are used; the bark is put in boiling water and given steam inhalation to remove insects from the tooth; the leaves are ground and eaten to remove worms in the stomach.

***Eclipta prostrata* (L.) L. (*E. alba* (L.) Hassk.) [Asteraceae]**

Karisalan kanni; prostrate or decumbent hairy herb; leaves and entire plant are used; these leaves with the leaves of neem, *Ocimum sanctum*, *Cassia leschenaultiana* are chewed and the juice is taken inside in empty stomach for anaemia, to inerease the secretion of the digestive juices, for smooth excretion of stools; the entire plant is ground and taken inside for jaundice, skin disease, dental diseases, ear and eye diseases, dropsy, disease of liver and spleen.

***Elettaria cardamomum* (L.) Maton var. *cardamomum* [Zingiberaceae]**

Elam (Wild cardamom); rhizomatous herb; NTFP.

***Eleusine coracana* (L.) Gaertn. [Poaceae]**

Keppai (Ragi); herb with stout culms; seeds; the seeds are powdered, mixed with warm water and applied on the forehead to remove heaviness of the head due to cough and cold and fever; gruel is made out of this and taken to remove body pain.

***Emilia sonchifolia* (L.) DC. [Asteraceae]**

Sulukku pattai thalai; herb with radical and cauline leaves; lower leaves lyrate, upper semiamplexicaul; bark and leaves are used; bark is ground with turmeric, fried in mud vessel and applied at the affected area for sprain and muscle catch; juice of leaves is applied for eye sores and night blindness.

***Entada pursaetha* DC. (*E. scandens* auct. non Benth.) [Mimosaceae]**

Thellikkai; unarmed large liana; NTFP.

***Erythrina variegata* L. [Fabaceae]**

Kattu murungai; medium-sized tree with smooth yellowish or greyish, shining bark; collected and used as firewood.

***Eucalyptus globulus* Labill. [Myrtaceae]**

Kungiliam; evergreen tree with bark peeling off as long thin strips; leaves ; leaves are boiled in water and taken bath to remove body pain and tiredness of the body; essential oil is extracted from the leaves and used externally for respiratory disorders.

***Eupatorium adenophorum* Spreng. [Asteraceae]**

Thiravida chedi; branched herb-stem granular; leaves; leaves are squeezed, the extract is mixed with lime powder and applied on the wound for cut injuries and to stop bleeding.

***Euphorbia hirta* L. (*E. pilulifera* Auct. non. L.) [Euphorbiaceae]**

Amma pacharisi thalai or pacharisi thalai or palkatti kirai; procumbent or erect herb; leaves; leaves are chewed in the mouth and the juice is taken in for dysentery and pain in the abdomen;

cooked as greens and eaten for general body health and to regulate mensus.

Euphorbia rosea **Retz. [Euphorbiaceae]**

Poolanthalai; Prostrate diffuse hairy herb; used for stall-feeding goats.

Euphorbia thymifolia **L. [Euphorbiaceae]**

Pachai pala chedi; prostrate hairy herb; leaves; the leaves are ground with turmeric and applied on the wounds for ulcers, blisters, boils in the genital area.

Euphorbia tirucalli **L. [Euphorbiaceae]**

Kalli; unarmed tall succulent shrub or small tree; latex; applied externally at the affected place to remove warts, applied in neuralgia and toothache, knee swelling, glandular swellings.

Evolvulus alsinoides **(L.) L. [Convolvulaceae]**

Ottu kodal thalai; prostrate silky-pubescent herb; entire plant; this plant is gently rubbed in the stomach and tied around the stomach for intestinal obstruction and hernia.

Ferula asafoetida **L. [Apiaceae]**

Perungayam; shrub or small tree; bark; this bark extract is used in combination with other extracts for stomach disorders and indigestion.

Ficus benjamina **L. [Moraceae]**

Malai-ichchi; Large tree without or with aerial roots; collected and used as firewood.

Ficus callosa **Willd. [Moraceae]**

Athi maram; tree; seeds; the seeds are taken out from the fruit, boiled and eaten for dysentery, motion with blood stains.

Ficus hispida **L. f. [Moraceae]**

Peiathai; hispid shrub or small tree; leaves; leaves are ground and applied on the wound in the tail of the horse for wounds in the tail of the horse; leaf extract is taken and given to pregnant women to induce labour pain and ensure safe delivery; the leaves are ground and applied at the centre of the head to induce labour pain and ensure safe delivery.

Ficus racemosa **L. (*F. glomerata* Roxb.) [Moraceae]**

Aththi; small or large tree with aerial roots; bark, trunk and root latex are used; since this tree gives out latex when cut, this is considered auspicious and the logs are used to put up pandals for auspicious occasions like marriage, village festival etc. as culture of the people; the root is ground, mixed with water and taken in for diarrhoea and diabetes; the latex is mixed with water and drunk for piles, diarrhoea and constipation.

Ficus religiosa **L. [Moraceae]**

Aarasa maram; large tree with or without aerial roots, often epiphytic; stem; as a preparation for marriage, first this stem is put in the centre of the thatched shed to be put up for the marriage; a tree with latex is considered to be auspicious and it is culturally important.

Ficus tsjahela **Rheede ex Burm. f. [Moraceae]**

Selai maram; large tree without aerial roots; bark rough, dark coloured; stem; the stem is used to make thatched sheds to keep the girls who have attained puberty in seclusion for fertility.

Flemingia macrophylla **(Willd.) Pr. ex Merr. (*F. congesta* Roxb. ex Ait.) (Fabaceae)**

Manjal Kamalai chedi; erect shrub with brownish tomentum; young shoot, leaves and root are used; the juice is squeezed out of this and applied on the body-head to foot-two leaves are stuck on the eyelids and taken bath for jaundice; root is ground and applied externally for ulcers and swellings.

Foeniculum vulgare **Mill. [Apiaceae]**

Sombu; aromatic herb with pinnatisect leaves; seeds; the seeds are soaked in water and eaten to reduce body heat, chest and kidney diseases.

Funeria hygrometrica **[Bryophyta]**

Bog moss; Gametophytic plants, consists of rhizoids, erect stem and leaves, stem often branched; NTFP.

***Girardinia diversifolia* (Link) Friis (*G. leschenaultiana* Decne.; *G. zeylanica* Decne.) [Urticaceae]**

Ocheri; stinging hairy herb; entire plant; the entire plant is uprooted and beaten on the body of the possessed person to drive away evil spirits from any possessed person.

***Gmelina arborea* Roxb. [Verbenaceae]**

Periya marudhu or Kumulam maram; deciduous unarmed tree with smooth whitish grey bark; wood and bark are used; wood is used to make furniture and to build houses; big log is taken, bored in the middle on both sides and the skin of goat or cow is tied on that to make musical instruments; bark is ground with salt and given inside for too much of loose motion; bark is ground and applied all over the body of children and given bath for healthy growth of children.

***Gmelina asiatica* L. [Verbenaceae]**

Siru kumul; armed, erect or scandent shrub; leaves; leaves are mixed with the leaves of *Solanum indicum*, cumin seeds and small onion, kept for one night and on the following morning distilled and drunk; after having taken the medicine, the patient should not step on chicken excreta or cow dung-should walk with slippers to cure STD, VD, to reduce body heat and to cure piles. These leaves are soaked in mud pot in rice cleaned water and drunk to reduce body heat and to cure piles.

***Grevillea robusta* A. Cunn. ex R. Br. [Proteaceae]**

Chavuku (Silver oak); tree; collected and used as firewood.

***Grewia tiliifolia* Vahl. [Tiliaceae]**

Thalai nar maram; medium-sized to large tree; bark; the bark is peeled out of the stem, soaked in water for one hour and used as shampoo for the hair to reduce the falling of hair and to increase the blackness of the hair; bark is dried and powdered and given in for dysentery.

***Gymnema sylvestre* (Retz.) R. Br. ex Schult. [Asclepiadaceae]**

Sakkarai kolli kodi; larges climbing shrub; leaves; the leaves are dried, ground and stored; one teaspoon of powder is taken every day before breakfast for forty days to reduce the level of sugar in diabetic patients; it temporarily arrests the sense of taste in a person who does not have diabetes.

***Helicteres isora* L. [Sterculiaceae]**

Singattam pattai or Asuvili maram; large shrub or small tree; bark grey, covered with stellate hairs when young; fruits, leaves and bark are used; fruits are eaten for intestinal complaints like colic, diarrhoea, chronic dysentery and to improve appetite; leaves are squeezed and the juice is taken inside for empyema, stomach affections, diabetes; bark is powdered, mixed with water, heated- when it becomes concentrated, applied on the affected part for paralysis, back pain, bone fracture, dislocation, cervical spondylosis; bark is powdered and taken inside for abdominal pain and dysentery.

***Hemidesmus indicus* (L.) Schult. var. *indicus* [Periplocaceae]**

Nannari; twining or prostrate or semi-erect laticiferous herb; root is used for stomach disorders, rheumatism, urinary diseases, skin troubles.

***Hemionitis arifolia* (Burm.) T. Moore [Hemionitidaceae]**

Kal Ilai; herb; leaves; these leaves are ground with the leaves of *Coleus aromaticus* and *Kleinia grandiflora* and the extract is applied all over the body three times a day for fainting attack and unsteadiness of the body.

***Heracleum rigens* Wall. ex DC. var. *rigens* [Apiaceae]**

Kattu kothamalli; sub erect herb with stout aromatic root; seeds; the seeds are ground and eaten; the seeds are used to make curry; for cough, cold and gastritis.

***Hibiscus cannabinus* L. [Malvaceae]**

Pulicha kirai; Unarmed herb; Cooked and eaten as greens.

***Hibiscus rosa-sinensis* L. [Malvaceae]**

Sembaruthi; shrub with ovate-lanceolate, serrate leaves; flower; the flowers are boiled and the extract drunk every day early morning in empty stomach to prevent heart problems, attacks and other ailments.

***Hydrocotyle javanica* Thunb. [Apiaceae]**

Meen kolingi thalai; stoloniferous creeping herb; leaves; the leaves are crushed with stones and spread on the stream and rivers to catch fishes.

***Impatiens balsamina* L. var. *balsamina* [Balsaminaceae]**

Karungiragam; succulent herb with elliptic-lanceolate, serrate leaves; petiole glandular; seeds and whole plant are used; these seeds are mixed with mustard seeds and palm jaggery, fried in gingeli oil and given inside to induce abortion; the whole plant is ground with crystal salt and applied externally, dried in sun and taken hot water bath for scabies.

***Ipomoea obscura* (L.) ker-Gawl. [Convolvulaceae]**

Sulukku thalai or Paal kodi or Pamparakan kodi; slender twiner (climber); leaves; the leaves are ground with small onion and applied at the place of sprain; the leaves are boiled and eaten; the leaves are squeezed out and the extract taken inside to ease out body pain; the leaves are ground with or without small onion and applied on the affected place for cut injuries and wounds due to dog bite.

***Ipomoea quamoclit* L. (*Quamoclit pinnata* (Desr.) Boj.) [Convolvulaceae]**

Vellakkodi; slender twiner; leaves and root are used; leaves are boiled and eaten for stomach disorders; leaves are dried, powdered and applied in hemorrhoids and also used in carbuncles; root is dried, powdered and used; powdered root is used as a sternutatory.

***Ixora nigricans* R. Br. ex Wt. & Arn. [Rubiaceae]**

Kadalai maram; shrub or small tree; young parts turning black on drying; collected and used as firewood.

***Jasminum cuspidatum* Rottl. [Oleaceae]**

Mullai kodi; erect shrub; entire plant or root; after delivery, if the placenta has not come out, then this plant is tied on the toe of the delivered mother to bring out the retained placenta.

***Justicia procumbens* L. [Acanthaceae]**

Kanal pachai ilai; slender herb; leaves; the leaves are ground with garlic, applied on the affected area of the skin in the evening and taken bath on the following morning; this is done for four days; this demands strict observance of diet-no dry fish, pumpkin, brinjal-to take only gruel; good for skin disease, dermatitis, scabies, eczema, icthyosis, skin plaques.

***Kalanchoe bhidei* Cooke [Crassulaceae]**

Ilai mulangi; Succulent herb; leaves; the leaves are mixed with the leaves of *Hemionitis arifolia*, *Kleinia grandiflora* and *Coleus aromaticus*, ground and applied on the centre of the head and all over the body three times for anaemia, myalgia, fainting attack and giddiness.

***Kalanchoe grandiflora* Wall. ex Wt. & Arn. [Crassulaceae]**

Malai pachai; herb; leaves; the leaves are ground and applied on the heel for pricking and burning sensation in the feet.

***Kleinia grandiflora* (Wall. ex DC.) Rani. (*Notonia grandiflora* Wall. ex DC.) [Asteraceae]**

Mosakkadhu ilai; succulent herb or undershrub; leaves; leaves are squeezed and the juice is applied on pimples; pepper water is made with these leaves and drunk for stomach ulcer.

***Lablab purpureus* (L.) Sweet var. *purpureus* [Fabaceae]**

Kattu mochai; climbing shrub; leaves and seeds are used; the leaves are squeezed in the palm, the extract is mixed with lime powder and applied externally on the throat; sprouted seeds are offered to deities during festival.

***Lagenaria siceraria* (Molina) Standley (*L. vulgaris* Seringe) [Cucurbitaceae]**

Soraikkai (bottle gourd); large climbing or trailing herb with bifid tendrils; leaves and fruits; boiled and eaten for swelling of the legs and hand and edema.

***Lantana camara* L. var. *aculeata* (L.) Moldenke [Verbenaceae]**

Karanai mullu; armed shrub with prickly branches; entire plant and leaves are used; this plant is gently rubbed from head to tail of a cow thrice at a time to get good milk yield from the cow; decoction is prepared with the leaves and given for tetanus, rhumatism, malaria.

***Lawsonia inermis* L. [Lythraceae]**

Maruthani; much branched shrub or small tree with grayish-brown bark; leaves; leaves are ground and applied on the heel, leg, palm and nails to reduce body heat and burning feet.

***Lepisanthes tetraphylla* (Vahl.) Radlk. [Sapindaceae]**

Nei kottan maram; tree; fruits; the fruits are collected and sold at Rs. 50/tin (about 5 kg).

***Leptadenia reticulata* (Retz.) Wt. & Arn. [Asclepiadaceae]**

Palakirai; twining shrub with yellowish, corky, deeply cracked bark; the leaves are cooked and eaten as greens.

***Leucas aspera* (Willd.) Link. (Lamiaceae)**

Thumbai kirai or Thumbai thalai; erect herb with diffuse quadrangular branches; leaves and flowers are used; cooked and eaten as greens to add essential vitamins to the body; the leaves of this plant, mixed with the leaves of neem tree and *Phyllanthus amarus*, dried and powdered-one tea spoon full of that powder is mixed with pure honey or ghee and taken inside daily to cure any skin ailment; these leaves are squeezed with small onion, salt and pepper and the drops are put in the nostrils to cure migrane head ache; these leaves are added with pepper and garlic and the extract is taken by squeezing them and that is poured into the ear for otrrhoea (C.S.O.M.) the flower of this plant is ground with garlic and pepper and taken inside to cure continuous fever; this flower is ground with garlic and small onion and applied externally for wounds and scabies.

***Leucas cephalotes* (Roth) Spreng. [Lamiaceae]**

Karunthumbai; erect bushy herb; leaves; leaves are squeezed in the palm with garlic and that extract is put in the nostrils for migrane.

***Leucas indica* (L.) R. Br. ex Vatke (*L. lavanduliifolia* J.E. Sm.) [Lamiaceae]**

Vilakkamaru chedi; slender puberulous herb; leaves; the leaves are ground with pepper and that mixture is put in a pure white cloth and the drops are let into the nostrils for ordinary head ache and migrane.

***Leucas vestita* Benth [Lamiaceae]**

Malai Thumbai; a tall twiny villous herb; leaves and entire plant are used; the leaves are ground with small onion and the extract is taken out; one drop is put in the nostril for migrane, heaviness of the head and to remove phlegm from the chest; when boiling rice or wheat, this plant is kept nearby; it is believed that the boiled grains will remain fertile and the yield and harvest will increase.

***Lindernia oppositifolia* (Retz.) Mukerjee (*Ilysanthes oppositifolia* [Retz.] Urb.) (*Bonnaya oppositifolia* (Retz.) Spreng.) (Scrophulariaceae)**

Kalamang kirai; erect marshy herb rooting at nodes; leaves; leaves are cooked and eaten for general body health.

***Lycianthes laevis* (Dural) Bitter. (*Solanum laeve* Dunal) [Solanaceae]**

Karantha kirai; herb; leaves; the leaves are chewed and the juice is taken in or the leaves are boiled and eaten for stomach ulcer; the leaves are cooked with monkey's meat, small onion, cumin seeds, pepper and a little salt-the extract is drunk and the cooked meat eaten for body pain and wheezing and asthma.

***Lycopersicon esculentum* Mill. (*L. lycopersicum* (L.) Karst.) [Solanaceae]**

Siru thakkali; aromatic herb; fruit and leaves are used; fruits are eaten raw; they are also boiled and eaten and used to make chutney and eaten for cooling the body; the juice is squeezed out and applied at the affected place for any infection in the face and skin plaques.

***Mahonia leschenaultii* (Wall. ex Wt. & Arn.) Takeda [Berberidaceae]**

Mulmurungai; tall shrub with persistent leaf bases on branchlets; collected and used as firewood.

***Malvastrum coromandelianum* (L.) Garcke [Malvacaeae]**

Siru Kurunthotti; annual herb or under shrub; leaves; leaves are ground and applied on the skin; decoction is made and given inside for dysentery.

***Mangifera indica* L. [Anacardiaceae]**

Maa maram (Mango tree); large tree with thick, rough, dark grey bark; leaves, fruit and seed are used; leaves are tied in a rope with neem leaves and that is tied in the village on any auspicious occasion especially during the village annual festival; raw fruit is ground and given to pregnant mothers for safe delivery; seed is dried, ground and given inside fro stomach disorders.

Melia azedarach **L. [Meliaceae]**

Malai vembu; Medium-sized tree with dark grey longitudinally furrowed bark; used as firewood.

Mentha arvensis **L. [Lamiaceae]**

Puthina kirai; an erect hairy or glabrous herb; leaves; leaves are ground with salt and eaten for constipation, indigestion and it is used as laxative.

Michelia nilagirica **Zenk. [Magnoliaceae]**

Shenbagam; tree or shrub; flower and bark are used; collected and sold as NTFP; the bark is boiled with arecanut and the mouth is gargled with that extract for swollen face due to bad tooth and sinus trouble.

Mimosa pudica **L. [Mimosaceae]**

Thotta chinungi; prickly woody herb with diffuse branches; leaves and root are used; leaves are crushed, juice taken and applied externally for sinus, sores and piles; the leaves are ground, applied at the affected area, dried and given warm water bath to prevent skin diseases; leaves are boiled with salt, kept soaked in that for one night and given bath with the leaves and water on the following morning for scabies and for the good growth of the body; decoction is made with the root and given in for gravel and other urinary complaints.

Moringa pterygosperma **Gaertn. (*M. oleifera* auct. non. Lam.) [Moringaceae]**

Murungai maram (drum stick); medium sized tree with thick, soft, corky, deeply fissured bark; leaves bark and flower are used; leaves are cooked and eaten as greens for general body health; the bark is ground with small onion and applied at the place of the dog bite to cure wounds due to dog bite; the flower is fried with ragi flour, wall sand and eaten for menorrhogia and overbleeding.

Mukia leiosperma **(Wt. & Arn.) Wt. (*Melothria perpusilla* Cogn.) [Cucurbitaceae]**

Uppilan kirai; climber; leaves; cooked and eaten as greens to add essential vitamins to the body.

Mukia maderaspatana **(L.) M. Roem. [*Melothria maderaspatana* (L.) Cogn.] [Cucurbitaceae]**

Mosa mosakkai; slender, prostrate or climbing herb; leaves; leaves are eaten as greens for general body health; these leaves are cooked with the leaves of white pumpkin and eaten for safe delivery and healthy growth of the child in the mother's womb given to pregnant mothers; these leaves are ground with coriander leaves and given inside to remove fainting sensation and giddiness.

Murraya koenigii **(L.) Spreng. [Rutaceae]**

Kariveppilai; aromatic shrub or small tree with dark brown or black bark; leaves, root and bark are used; leaves are ground and given in for diarrhoea, dysentery and for checking vomiting; juice from the root is taken and given in for relief from ranal pains; the bark is ground with garlic, pepper and coriander, everthing is boiled and distilled and taken in for body pain and gastritis; the bark or small stick is ground and applied at the place of the boils in eyelids.

Musa paradisiaca **L. [Musaceae]**

Vazhai (banana); stoloniferous plant with cylindric trunk; fruit, sap and bark (country variety) are used; fruit is eaten for laxative, intestinal disorder, uraemia; camphour is put inside the banana fruit and eaten to induce abortion; unripe banana is burnt, dried in dew and eaten in empty stomach for dysentery, for pain in the genitals after urinating, VD and STD; the bark is heated a little and the extract is poured into the ear for pain.

Myristica fragrans **Houtt. [Myristicaceae]**

Jathikai; aromatic tree with greyish-black, longitudinally fissured bark; fruit; the fruit is dried, powdered along with gallnut and gooseberry and one teaspoon of medicine is taken every day before going to bed at night for migrane head ache, any other head ache, worm infestation and liver diseases.

Naravelia zeylanica **DC. [Ranunculaceae]**

Theenikirai; climbing shrub; cooked and eaten as greens.

Ocimum americanum **L. (*O. canum* Sims.) [Lamiaceae]**

Kanjan korai; herb with elliptic-oblong, serrate leaves; leaves; leaves are ground with small onion and the extract is given inside for whooping cough for children, dyspnoea, lower respiratory infection, ordinary cough, digestive disorders, fever; these leaves are boiled with the leaves of Eucalyptus and lime and steam inhalation is given with that for too much of head ache.

Ocimum gratissimum L. [Lamiaceae]

Pachai Ilai or Thiru neetru pachai; erect, aromatic under shrub; leaves, seeds and entire plant are used; the extract of the leaves is taken by squeezing them in the palm and applied at the affected area for dermatitis, skin diseases, eczema, icathyosis, acne, vulgaris, skin plagues, ear ache, tooth ache; the leaves are ground and applied on the feet for burning feet.; the extract from the leaves is given inside especially for children for abdominal colic, for cough, cold, to get the chest phlegm out, stomach disorders; seeds are crushed and given inside for headache, neuralgia, dysentery; the entire plant is ground and given inside for swelling, abscess, cough and arthritis.

Ocimum tenuiflorum L. (_O. sanctum_ L.) [Lamiaceae]

Thulasi; aromatic pubescent herb; leaves, stem and seed are used; leaves are boiled with cumin seeds, garlic, dry ginger, pepper and palmyra jaggery and that decoction is drunk for T.B., whooping cough, asthma, bronchitis and chest pain; these leaves are added with leaves of *Adhatoda zeylanica* in equal proportion, dried, powdered, mixed with honey and eaten for asthma; these leaves are ground with pepper and a drop of water is added and that is distilled and a drop is put in the nostril for head ache; the leaves, stem and seeds are boiled in water with other ingredients and distilled and drunk (at least thrice in a day) for fever, mental disorders, digestive disorders, cough, ulcers, leucorrhoea, head ache and body pain.

Opilia amentacea Roxb. [Opiliaceae]

Theenili kirai; climbing shrub; leaves; the leaves are boiled with cumin seeds and the decoction is drunk for edema, good for pregnant mothers.

Orthosiphon aristatus (Bl.) Miq. (_O. stamineus_ Benth.) [Lamiaceae]

Karum thulasi or Kattu chemarai; erect herb; leaves; the leaves are ground with paddy and given inside after the last day of mensus and this will aid in conception; decoction of the leaves is given inside for kidney and bladdar diseases; the leaves are mixed with small onion, ginger, cumin seeds, monkey's meat and leaves of *Melothria maderaspatana*–all boiled together and the extract is drunk for three days for T.B., fever, heart problems, to strengthen bones and nerves; leaf juice is mixed with lime juice and applied on pimples in the face; leaves are ground with small onion, applied externally at the affected place and dried for scabies and boils ; leaves are ground and given in for stomach ache.

***Oryza sativa* L. [Poaceae]**

Nellu (paddy); annual or perennial grass; staple food for tribals.

***Osyris arborea* Salzm. ex Decne. var. *puberula* (Hook. f.) Kumari in Biol. [Santalaceae]**

Sundari villi maram; small shrub; bark; the bark is used instead of arecanut to chew beatle leaves for better digestion.

***Oxalis corniculata* L. [Oxalidaceae]**

Chengarai kirai; diffuse creeping herb, rooting at nodes; cooked and eaten as greens.

***Oxalis latifolia* H.B. & K. [Oxalidaceae]**

Yanai pulinjarai; bulbous herb; leaves; boiled and eaten as food for general body health; fresh juice from the leaves is taken and given inside for fever, head ache, abdominal disorder, venereal diseases and digestive disorders.

***Punica granatum* L. [Punicaceae]**

Mathulai (Pomegranate); armed shrub or small tree; bark, unripe young fruits and ripe fruits are used; bark is ground and given inside to expel tape-worm; the young fruit is ground and given inside; the young fruit is ground, added in goat's milk and taken in for dysentery; fruits are collected and eaten for dysentery and other stomach disorders.

***Papaver somniferum* L. [Papaveraceae]**

Kasakasa; erect unbranched herb; seed; the seed is used to make extract with other ingredients for common ailments.

***Passiflora calcarata* Mast. [Passifloraceae]**

Thatpoot kodi; wiry climber; entire plant; NTFP.

***Passiflora edulis* Sims [Passifloraceae]**

Passion fruit; shrub; NTFP

***Pennisetum typhoides* (Burm. f.) Stapf et Hubb. [Poaceae]**

Kambu (Millet); erect herb with slender stems; used for worship of deities during the annual festival.

***Pentatropis capensis* (L.f.) Bullock (*P. microphylla* (Heyne ex Roth) Wt. & Arn.) [Asclepiadaceae]**

Uppilan kirai; climbing shrub; cooked and eaten as greens.

***Pergularia daemia* (Forssk.) Chiov. (*Daemia extensa* (Jacq.) R. Br.); (*Pergularia extensa* (Jacq.) N.E. Br.) [Asclepiadaceae]**

Veliparuthi; hispid climber; leaves and entire plants are used; leaves are ground and given inside for children affected by evil spirits; leaves are boiled and eaten to remove tiredness of the body; leaf decoction is taken and given for asthma; leaves are crused and the juice is given inside and also applied externally for infantile diarrhoea, combined with lime applied to rheumatic swellings; extract of the plant is taken and given inside for uterine and menstrual troubles and to facilitate parturition.

***Phaseolus calcaratus* Roxb. [Fabaceae]**

Beans; slender hairy climber; used as food.

***Phoenix loureirii* Kunth (*P. humilis* Royle var. *pedunculata* Becc.) [Arecaceae]**

Malai Icham; palm with very short stem; NTFP; the pith and young shoots are eaten.

***Phoenix sylvestris* (L.) Roxb. [Arecaceae]**

Icham (Date-palm); tree; NTFP.

***Phyllanthus amarus* Schum. & Thonn. (*P. niruri* auct. Non L.) [Euphorbiaceae]**

Kizha nalli or Kila Kodi; slender herb; entire plant and leaves are used; the entire plant is uprooted and tied at the ankle of pregnant mothers for safe and smooth delivery; leaves are boiled and eaten as greens for general body health.

Phyllanthus debilis **Klein ex Willd. [Euphorbiaceae]**

Perung kila nalli or peru ilai; slender herb; leaves used; leaves are ground, mixed with goat's milk and drunk early morning in empty stomach for jaundice and TB.

Phyllanthus emblica **L. (*Emblica officinalis* Gaertn.) [Euphorbiaceae]**

Nelli; small tree with hairy reddish-brown branchlets; fruits, leaves and bark are used; raw fruits are eaten for cooling, laxative, a rich source of Vitamin C, constipation, fainting, anaemia, piles, asthma, hypertension, tuberculosis, cardiac disease, chronic fever; fruits are collected and sold as NTFP-peak season October to January; fruits are collected and used to prepare pickle; leaves or fruits are heated in coconut oil and applied on the hair to remove dandruff from hair; bark is boiled and used for steam inhalation to cure insect or worm infested tooth.

Pinus roxburghii **Sarg. (*P. longifolia* Roxb.) [Pinaceae]**

Theekuchi maram; tall tree with whorled branches; stem and leaves are used; the stem is used to make musical instruments like flute; leaves are boiled with cumin seeds and small onion and the extract is taken for three days to cure VD and STD.

Piper betle **L. [Piperaceae]**

Vetrilai; woody climber with adventitious root; stem of the betle leaf, leaves and root are used; the stem is mixed with gallnut, *Alpinia calcarala, Acorus calamus, Ferula azafoetida*-ground and mixed in mother's milk and given inside for digestion for babies; leaf extract is taken and given inside, decoction of leaves is also given inside for cough, cold, any other bronchial and respiratory problems, healing wounds; root is ground with black pepper and given inside to produce sterility in women.

Piper mullesua **Buch.-Ham. ex D.Don (*P. brachystachyum* Wall. ex Hk.f.) [Piperaceae]**

Milagu; climber with slender branches; root, seed and wild seed are used; root is ground with clove and applied on the affected tooth to cure tooth ache; ground with salt and applied on the forehead for head ache; soaked with honey and eaten to get good apitite; collected and sold at Rs.5/kg. The peak season is April and May.

***Piper nigrum* L. [Piperaceae]**

Milagu (Pepper); much branched climbing shrub rooting at nodes; NTFP.

***Piper trioicum* Roxb. (*P. attenuatum* Buch.- Ham. ex Miq.; *P. bantamense* Bl.) [Piperaceae]**

Kaattumilagu keerai; slender climber with stiff branches; cooked and eaten as greens.

***Plectranthus amboinicus* (Lour.) Spreng (*Coleus amboinicus* Lour.; *C. aromaticus* Benth.) [Lamiaceae]**

Semarai; large succulent herb with aromatic leaves, flowers small, pale purple; etire plant and leaves are used; the entire plant is tied at the entrance of the house on the roof to ward off attack by evil spirits; the juice is taken from the plant and given inside for urinal disorders, vaginal discharges, snake bite; the leaves are squeezed in the palm and the extract is given inside in empty stomach for cholera and diarrhoea. This is given to small children for intestinal colic, common cold, cough, whooping cold and indigestion; this leaf extract is added with salt and given inside. Leaf decoction is also prepared and given for chronic cold, asthma, stomach pain and loose motion.

***Plectranthus barbatus* Andr. (*Coleus barbatus* (Andr.) Benth; *C. forskohlii* (Willd.) Briq.) [Lamiaceae]**

Parai pidungi; sparsely branched rhizomatous herb with stout, cylindric, woody stem, villous with long hairs; root and whole plant are used; the root is ground and applied on the chest area where there is pain for chest pain and gastritis; the whole plant is ground and given inside for hemorrhage, worm infestation, edema, piles, cough.

***Plectranthus caninus* Roth. [Lamiaceae]**

Naga thali chedi; fleshy herb; root used; the root is chewed and the extract gulped. The molass is applied at the spot of the bite for snake bite or any poisonous bite.

***Plectranthus malabaricus* (Benth) Rott. Willemse (Lamiaceae)**

Muthupattai; a tall tree; leaves and entire plant are used; leaves are folded and tied on the wound in the tail of the horse; this is ground with horse urine and a little salt is added and applied on scabies.

Polytrichum densifolium **[Bryophyta]**

Rock Moss; gametophytic plant, consists of rhizoids, underground rhizome, erect stem and leaves. The young leaves at the apex of a stem are arranged in a definite manner. The arrangement is directly correlated with the number of cutting faces of the apical cell of the stem; NTFP.

Polygala arillata **Buch. Ham. ex D.Don. [Polygalaceae]**

Pinagari; erect shrub; collected and used as firewood.

Polygala arvensis **Willd. (*P. chinensis* auct. non. L. 1753) [Polygalaceae]**

Siriya nangai; diffuse herb; leaves used; leaves are eaten raw (If poison has entered the body, the leaves will be sweet. If poison has not entered the body, the leaves will be bitter) to remove poison in the body.

Polygonum chinense **L. [Polygonaceae]**

Chokkattan karumbu; a rambling undershrub with white or pink flowers; leaves with the stem and leaves are used; leaves are chewed and the juice is taken in. The skin of the stem in peeled off and that stem is chewed to stop leucorrhoea; the extract from the leaves is taken and poured into the ear for pricking pain in the ear.

Pongamia pinnata **(L.) Pierre (*P. glabra* Vent.) [Fabaceae]**

Ponga maram; medium-sized tree with greyish-green or brown, smooth or tuberculate bark; leaves, root, bark and seed are used; leaf juice is taken out and given for breathing troubles, any bronchial problems, flatulence, dyspepsia, diarrhoea, cough, gonorrhoea, leprosy; root is ground, juice taken and given, applied externally for cleaning foul ulcers, fistulous sores, cleaning teeth and strengthening gums; bark is powdered and given and bark decoction is also given for piles; fatty oil is taken out from the seeds and used externally for scabies, herpes, leucoderma and other cutaneous diseases.

Portulaca oleracea **L. var. *oleracea* [Portulacaceae]**

Kolikal kirai; prostrate or erect succulent herb with purplish or green stems; cooked and eaten as greens.

***Portulaca pilosa* L. ssp. pilosa (*P. tuberosa* Roxb.) [Portulacaceae]**

Kilavan thadi chedi; herb with tuberous root; root used; the root is chewed, the extract is taken inside. The chewed remains are applied at the mouth of the bite to remove poison due to some poisonous bite.

***Portulaca quadrifida* L. [Portulacaceae]**

Kolikkal keerai or pasali keerai; slender, diffuse, succulent, annual herb, rooting at nodes with suberect branches; leaves and whole plant are used; leaves are cooked with green gram, small onion and salt and eaten to aid conception; entire plant is boiled and eaten for piles, abdominal disorders, vomiting, asthma, ulcers, inflammation.

***Psidium guajava* L. [Myrtaceae]**

Koyya; small tree or large shrub; bark, leaves, young unnipe fruit and fruits are used; decoction of the bark is made and given for diarrhoea; young leaves are eaten raw for dysentery; young unripe fruits are mixed with fenugreek and taken for dysentery; fruits are eaten for cooling, laxative, useful in colic and bleeding gums.

***Psilotrichum scleranthum* Thwaites [Amaranthaceae]**

Nayurunji chedi; shrub, branches tomentose; flowers white; leaves used; leaves are ground and applied at the place of dog bite at once.

***Pterocarpus marsupium* Roxb. [Fabaceae]**

Vengai maram; medium-sized to large deciduous tree; fruits and latex are used; fruits are collected and sold at Rs. 2 per kg. The peak season for the collection is from October to November; NTFP; the latex is applied to joints for joint pain and mixed with hot water and drunk for chest pain and heart pain.

***Raphanus sativus* L. [Brassicaceae]**

Mullangi (raddish); erect herb with napiform root; root used; root is boiled / cooked and eaten for blood purification, prevents the formation of kidney stone and cures stomach disorders.

***Ricinus communis* L. [Euphorbiaceae]**

Aamanakku; erect shrub or small tree; root and leaves are used; decoction is prepared with this root and given inside for lumbago; poultice of leaves is applied externally for boils and sores; castor oil is applied to the leaves and heated a little and tied on the boils to break and get cured.

***Rubia cordifolia* L. [Rubiaceae]**

Thondai Aruppan kodi; scabrid climber with quadrangular branches; leaves used; leaf decoction is given for vermifuge; supposing a boy or a girl (a man and a woman) stealthily talk in the forest-if they are possessed by evil spirits, these leaves are ground and applied from head to foot-it is believed that the evil spirits will run away.

***Rubus niveus* Thunb. [Rosaceae]**

Seetipalam or Karun cheendi chedi; shrub; fruit, root and young shoot are used; fruits are collected and eaten as food; the extract from the root or young shoots is squeezed out, mixed with goat's milk and given orally for mouth ulcers.

***Ruta chalepensis* L. (*R. graveolens* L. var. *angustifolia* sensu Hk. f.) [Rutaceae]**

Aruvatham pachai or Punithalai; aromatic herb / undershrub; leaves used; these leaves are ground with the leaves of lemon tree, acorus, and garlic, applied on the body of small children and given bath for general health of children especially to cure some skin ailments; these leaves are ground and applied all over the body especially at the affected place, dried and taken / given bath after half an hour in hot water for dermatitis, scabies, eczema, swelling or tumour; the bark removed stem is cut into small pieces, joined in a thread and tied at the wrist of small children to ward off evil spirits; the leaves are ground and applied on the head of children to prevent children being affected by evil spirits.

***Sansevieria roxburghiana* Schult. & Schult. f. [Liliaceae]**

Marulu kai chedi; herbs with creeping rhizome; tender shoots, root and fruits are used; juice of tender shoots is given to children for cleaning the phlegm from the throat; root is dried, powdered and given in for constipation, piles, cough, asthma, tuberculosis and fever; fruits are spread on the bed of the girl who has attained puberty when she is kept in isolation and when they go out for natural calls, they have to carry this fruit to ward off evil forces and spirits; fruits are squeezed in the ear for ear pain.

***Sapindus emarginatus* Vahl. [Sapindaceae]**

Poosakkai maram; tree; fruit used; fruits are collected and sold as NTFP for Rs. 5/kg; NTFP.

***Schefflera stellata* (Gaertn.) Harms [Araliaceae]**

Kurangu kai maram; climbing shrub; bark used; the bark is ground with turmeric and taken inside for cholera.

***Sechium edule* (Jacq.) Sw. [Cucurbitaceae]**

Chav chav; climber; commonly used vegetable for food.

***Senecio scandens* Buch.-Ham. ex D. Don. [Asteraceae]**

Kodivili ver; slender climber; root used; the root is ground and applied on the head and forehead for migrane head ache and heavy head ache.

***Senna alexandrina* Gars. ex. Miller [Caesalpiniaceae]**

Ponnaavarai; under shrub; root used; the root is burnt in fire, powdered and given inside for indigestion, bloated abdomen and intestinal obstruction.

***Senna auriculata* (L.) Roxb. (*Cassia auriculata* L.) [Caesalpiniaceae]**

Avarai chedi; tall shrub with smooth reddish-brown bark and foliaceous stipules; seeds in flower used; the seeds in the flower are added with coffee and drunk to reduce body heat.

***Senna occidentalis* (L.) Link (*Cassia occidentalis* L.) [Caesalpiniaceae]**

Vatta thagarai; diffuse foetid undershrub; leaves and seeds are used; leaves can be cooked and eaten as greens. The leaves are ground and applied on the forehead for headache and to remove phlegm from the chest (lungs); seeds are ground and applied externally for skin diseases.

***Senna tora* (L.) Roxb. (*Cassia tora* L.) [Caesalpiniaceae]**

Thagara chedi or vatta thagara kirai; foetid smelling herb; root used; the root is burnt and boiled with water and that decoction is distilled and given inside for indigestion, bloated abdomen, intestinal obstruction, skin diseases and constipation.

***Sesamum indicum* L. (*S. orientale* auct. non L.) [Pedaliaceae]**

Yellu; erect herb; seed used; seed paste is prepared and applied to piles; oil is extracted from the seed and that gingelly oil, mixed with palmera extract is given in to induce abortion.

***Sesbania grandiflora* (L.) Pers. [Fabaceae]**

Agathi kirai; medium-sized tree; leaves, root, bark and flower are used; leaves are cooked and eaten. Decoction is made out of the leaves and drunk for general body health. Decoction is also given for diarrhoea and dysentery, given as an expectorant; juice of the root is mixed with honey and given in for scabies, ulceration of tongue and alimentary canal; bark is powdered and applied externally on the stomach for diarrhoea and dysentery; bark decoction is made and given in to improve vision; flower is crushed, juice taken and drops are put in the eyes.

***Sida acuta* Burm. f. [Malvaceae]**

Malai thangi ver; erect herb; root used; the root is dried and powdered and given inside for diarrhoea, worm infestation, arthritis.

***Sida cordata* (Burn.f.) Borss. (*S. veronicifolia* Lam.) [Malvaceae]**

Velakaditha chedi or palam pasi chedi; slender trailing herb with sparsely pubescent or slightly viscid branchlets; leaves or fruits are used; the extract from the leaves is given inside for menorrhagia; the leaves are boiled and the decoction is drunk for the healthy functioning of the heart; the leaves are ground, mixed with the white of the egg and applied on the leg or head for three days to reduce swelling in the hands and legs-edema and burning feet; the leaves are ground and applied on the forehead for head ache; leaves are given to pregnant women to stop diarrhoea; the leaves or fruits are cooked and eaten for general body health.

***Sida glutinosa* Comm. ex Cav. [Malvaceae]**

Ottu pakkutti thalai; under shrub; leaves used; leaves are ground with lime powder and applied on the wound for cut injuries.

***Sida rhombifolia* L. ssp. *retusa* (L.) Borss [Malvaceae]**

Kurun thatti; woody herb with stellate-tomentose braches; leaves used; the leaves are ground and mixed with goat's milk and given inside for burning feet and hands.

***Solanum anguivi* Lam. var. *auguivi*. (*S. indicum* anct. non. L. 1753) [Solanaceae]**

Chitarangai chedi; armed stellate-tomentose shrub; fruit, root and leaves are used; the fruits are pounded with tamarind and salt and eaten as side dish for food for any worm infection and to remove the worms in the intestines; the root is boiled with *Strobilanthes kunthiana* flower, ginger, pepper, small onion and palm jaggery and that decoction is drunk for general body health; the leaves are mixed with the leaves of *Gmelina asiatica*, cumin seeds, small onion and this mixure is soaked in rice boiled water for one night and this water is distilled and drunk to cure STD, VD and other such sex related diseases.

***Solanum erianthum* D.Don. (*S. verbascifolium* auct. non. L. 1753) [Solanaceae]**

Valaran katti thalai or Pisilan thalai; densely stellate-tomentose large shrub; leaves and root are used; the leaves are ground and applied on the wounds in the tail of horses for cut injuries and other wounds; the leaves are ground with salt and applied all over the leg to prevent leech bite; leaf extract is given in for vaginal discharge and to ease body pain, vertigo and urinary troubles.

***Solanum giganteum* Jacq. [Solanaceae]**

Sama sunda thalai; white tomentose armed shrub with tomentum except on the leaves above; leaves used; leaves are used for dressing wounds and foul ulcer; the extract of the leaves is taken and given inside for stomach disorders and diarrhoea.

***Solanum melongena* L. var. *melongena* [Solanaceae]**

Kaththiri; herb; commonly used vegetable for food.

***Solanum nigrum* L. [Solanaceae]**

Milagu thakkali; unarmed herb; leaves used; the leaves are eaten raw or cooked and eaten for stomach ulcer.

***Solanum virginianum* L. (*S. xanthocarpum* Schard. & Wendl.) [Solanaceae]**

Kandankathiri; diffuse spiny herb; seeds, leaves and fruits are used; seeds are taken out of the fruits and applied externally on the teeth for tooth ache; juice of the leaves is mixed with black pepper and given inside for rheumatism; the fruits are dried, ground and applied at the affected tooth for tooth ache; fresh fruits are squeezed and the juice is applied on the leg to prevent leech bite.

Solanum torvum **Sw. [Solanaceae]**

Sundakkai; armed-hairy shrub; leaves, fruits and root are used; the leaves are ground and tied on the tail of the horse for wounds in the tail of the horse; leaves are used for making decoction and that is given for cough; after removing the seeds from the fruit, they are joined together in a thread and tied around the neck of small children. As the fruits dry up, cough also will decrease. The dried fruits will automatically fall down. This is done for whooping cough, lower respiratory infection, and regurgitation of milk by small children; the fruits are used to prepare curry ofter removing the seeds for diarrhoea and worm infestation; root is dried, ground and applied externally to cracks in the feet.

Solanum tuberosum **L. [Solanaceae]**

Urulai kilangu (Potato); branched bushy tuberous herb; commonly used vegetable for food.

Sorghum vulgare **L. [Poaceae]**

Cholam (Maize); tall grass with long, lanceolate leaves; used for worship of deities during the annual festival.

Spilanthes calva **DC. (*S. acmella* auct. non (L.) Murr. 1844) [Asteraceae]**

Norupan chedi; erect herb; flowers and leaves are used; flowers are chewed for throat affections and used in stammering of children; decoction of the leaves is prepared and given inside for dysentery, scabies; these leaves are ground with stream crabs, boiled and that heat is applied all over the body for osteo arthritis and edema.

Stenosiphonium russelianum **Nees [Acanthaceae]**

Periya Kurinji; erect herb; leaves used; decoction is prepared with these leaves, pepper, dry ginger and garlic and taken inside for back pain and any problem in the spinal cord.

Stephania hernandifolia **Walp. [Menispermaceae]**

Vatta kanni thalai; climbing herb; leaves used; the extract from the leaves is taken and given inside for stomach disorders and diahorrea.

Strobilanthes kunthiana **(Nees) T. Anderson ex Benth. [Acanthaceae]**

Kurunji; shrub; leaves used; decoction is prepared with pepper, dry ginger, garlic-distilled and taken inside for back pain, pain in the spinal cord and cervical spondylosis; this is culturally significant plant. Supposing a young boy and a girl have eloped, this practice is done. Two leaves of this plant are taken, ground yam and turmeric is put on these leaves. The leaves are let to float on the stream of the river by a person who has taken bath in the same. If both the leaves join now and then, then it is believed that the eloped couple will live together safely. If both the leaves float separately, then, it is believed that the eloped couple will separate. The leaves are used to stall-feed the goats.

Stylosanthes fruticosa **(Retz.) Alst. (*S. mueronata* Willd.) [Fabaceae]**

Siru thumbai; decumbent herb; leaves used; the leaves are ground and applied on the forehead for head ache.

Styrax benzoin **Dryander [Styraceae]**

Samburani maram; a small or fairly large tree, bark rough with deep vertical tissures greyish-brown, woodred-makes good gunpowder-charcoal, the twigs are woven into baskets; NTFP.

Syzygium aromaticum **(L.) Merr. & Perry. (*Eugenia caryophyllata* Thunb.) [Myrtaceae]**

Kirambu (clove); tall, erect, pyramidal or conical tree with thin, grey smooth bark; clove used; the clove is ground and put at the place of decayed tooth.

Syzygium cumini **(L.) Skeels (*S. jambolanum* (Lam.) DC.) [Myrtaceae]**

Naval maram; large tree with greyish-brown thick bark, exfoliating in woody scales; bark and seeds are used; bark is soaked, ground and taken inside for dysentery; decoction is made with the bark and drunk for diabetes; seeds are powdered and taken for diabetes.

Tabernaemontana gamblei **Subram & Henry [Apocynaceae]**

Andhi vattam poo; shrub; flower used; the flower is soaked in fresh mother's milk-ground with mother's milk and distilled and poured in drops in the affected eye for pain in the eyes; ground turmeric is mixed with water-kept in a vessel-this flower and lemon grass are added with that water-that container is continuously stricken. The affected person is made to concentrate and look inside that vessel to remove blood clot in the eyes and to remove white reflex in the cornea.

Talinum portulacifolium **(Forsk.) Asch. [Portulacaceae]**

Paruppukkirai; under shrub; leaves used; the leaves are cooked and eaten as greens and this adds essential vitamins to the body.

Tamarindus indica **L. [Caesalpinaceae]**

Puli; large tree with brownish or grey longitudinally and horizontally fissured bark; leaves used; leaves are boiled in a covered vessel and the face washed with that water for eye pain.

Tephrosia fusca **Wight & Arn. [Fabaceae]**

Meen kolinji; shrub; entire plant used; this plant is ground with the sand of anthill and sprayed at the place where there are plenty of fishes in the stream. The fishes faint and float on the surface of the stream which could be collected.

Tephrosia tinctoria **Pers. [Fabaceae]**

Karung kolingi; under shrub; seed used; the seeds are boiled with sugar and cow's milk and given after the third day of mensus upto seven days to help in conception.

Tephrosia purpurea **Pers. [Fabaceae]**

Kolingi; much branched erect glabrescent undershrub; root, pod and seed used; root is chewed and the extract taken in for too much of stomach ache; root is boiled and used to give steam inhalation for general body health and to cure heaviness of the head; decoction is made out of the root and given for dyspepsia, diarrhoea, rheumatism, asthma, urinary disorders; a liniment is prepared from the root and used in elephantiasis, relief from asthma and cough; pulverised root is smoked; decoction of the pod is made and drunk as a verimifuge and vomiting; oil is crushed out of the seed and used externally at the affected place for scabies, itch, eczema, other skin eruptions.

Terminalia arjuna **(Roxb. ex. DC.) Wt. & Arn. [Combretaceae]**

Marudham; large tree with white grey or pinkish-green, smooth bark, exfoliating in large, thin, irregular sheets; fruit used; fruits are collected and sold at Rs.8/kg for NTFP.

***Terminalia chebula* Retz. [Combretaceae]**

Kadukai (gallnut); large tree with dark-brown, cracked bark, exfoliating in woody scales; fruit used; the fruit is ground with salt and applied at the affected tooth for tooth ache, dentrifices; the fruit is powdered and smoked for asthma; the seed is removed from the fruit, ground, boiled, kept soaked in the same for one night, distilled and drunk on the following morning for dysentery; the fruits are collected and marketed as NTFP.

***Tetrastigma lanceolarium* (Roxb.) Planch. [Vitaceae]**

Iaiviral kirai; climbing shrub; cooked and eaten as greens.

***Themeda cymbaria* Hackel [Poaceae]**

Tharuvaipul; slender herb; used to thatch houses.

***Themeda tremula* (Nees ex Stead.) Hack [Poaceae]**

Pullu; herb; used to thatch houses.

***Themeda triandra* Forsskal [Poaceae]**

Pothaipullu; herb; used to thatch houses.

***Thespesia populnea* (L.) Sol. ex Correa [Malvaceae]**

Poovarasa maram; large shrub or small tree; bark grey-brown, fissured, often scaly; leaves used; the leaves are ground with small onion and salt, applied all over the head, allowed to dry and taken bath afterwards to blacken hair and to reduce falling of hair.

***Thunbergia alata* Boj. ex Sims. [Thanbergiaceae]**

Vettu kaya thalai; slender twiner; leaves used; these leaves are ground with the leaves of *Desmodium trifolium* and given inside for general body health; the leaves are ground and applied outside for strengthening the body and applied on the head for relief from head ache.

***Toddalia asiatica* (L.) Lam [Rutaceae]**

Perugala thalai; aromatic, prickly, scandent shrub; leaves and entire plant are used; after the birth of the baby, these leaves are ground and applied all over the body to prevent the attack from evil spirits; leaves are boiled and lime or lemon juice is added to that and steam inhalation given for body pain and myalgia; leaves are cooked with wheat flour and eaten for continuous cough and chest pain;

this plant is carried when coming back after attending any funeral. This is also applied on children to ward off the attack from evil spirits.

***Toona ciliata* Roem. var. *ciliata* (*Cedrela toona* Roxb. ex Rottl. & Willd.) [Meliaceae]**

Santhana vembu (Indian mahogany); tall deciduous tree; used to make household articles.

***Trachyspermum ammi* (L.) Sprague. (*Carum copticum* Benth. ex Hiern.) [Apiaceae]**

Omum; erect branched herb; leaves used; the leaves are ground with the leaves of *Cicer arietinum* and mustard seed and taken inside to induce abortion.

***Trichosanthes lobata* Roxb. [Cucurbitaceae]**

Val sorai; climber; fruit and leaves are used; fruit is cooked and eaten to reduce body weight., to cure thyroid problem; leaves are cooked and eaten for wheezing and dyspnoea.

***Tridax procumbens* L. [Asteraceae]**

Chani pottu thalai; procumbent herb; leaves used; the leaves are squeezed in the palm and the extract is administered orally. This has to be given daily for three months. The patient has to lick seven drops from the palm and seven drops from the back of the palm. Then the remaining juice is taken inside. This is given only to women and they should avoid sexual intercourse for three months to prevent pregnancy permanently; the extract from the leaves is applied for ordinary wounds.

***Trigonella foenum-graecum* L. [Fabaceae]**

Vendyam; aromatic herb; seeds used; these seeds are ground with cumin seeds and salt, mixed with water and given inside for dysentery and diarrhoea.

***Triumfetta rotundifolia* Lam. [Tiliaceae]**

Thuthi chedi or Ottuka chedi; woody herb; young shoots and leaves used; four fresh leaves with seven garlic cloves are chewed in the mouth and the breath is blown into the ear of the affected person. These chewed leaves are to be thrown after rotating it around the head of the affected person thrice for gastritis; leaves are ground with fenugreek and taken inside for pain in the abdomen and dysentery.

***Typhonium trilobatum* Schott. [Araceae]**

Karunai kilangu; herb; yam used; the yam is dug out, boiled and eaten. It is good for body health especially for stomach disorders.

***Vernonia anthelmintica* (L.) Willd. (*Centratherum anthelminticum* [L.] Ktze.) (Asteraceae)**

Kattu chiragam (Cumin seeds); erect herb. Leaves, elliptic-lanceolate; NTFP.

***Vigna mungo* (L.) Hepper (*Phaseolus mungo* L.) [Fabaceae]**

Vulundhu (black gram); erect hairy herb; seed (pulse) used; the pulse is ground with the white of egg and leaves of *Blepharis boerhavifolia* and applied at the place of the fracture to set right bone fracture and dislocation.

***Vigna unguiculata* (L.) Walp. ssp. *unguiculata* (*Dolichos biflorus* L.) [Fabaceae]**

Kollu / Kaanam (Cowpea); suberect herb; seeds used; the seeds are ground with salt, boiled and applied externally at the affected place for swelling in the joints and edema.

***Vitex negundo* L. [Verbenaceae]**

Nochi thalai; large shrub or small tree; leaves and flowers are used; leaves are mixed with the leaves of *Ocimum acorum* and crystal salt, ground and taken in for cough and cold; leaves are ground and that paste is applied on the forehead or leaves are boiled and the steam is used for inhalation for head ache; flowers are boiled and that decoction is drunk for diarrhoea, fever and liver complaints.

***Vitex trifolia* L. [Verbenaceae]**

Anjalai nochi; aromatic large shrub or small tree with pale grey, smooth bark; leaves, flower and fruits are used; young leaves are squeezed with a few drops of water, distilled with a nice cloth and if there is head ache on the left side of the head, then a few drops are put in the right nostril. If there is head ache on the right side, then a few drops are put on the left nostril for migrane head ache; the extract from the leaves is given inside for tuberculosis; flowers are soaked in honey and eaten for fever accompanied by vomiting, severe thirst; fruits are eaten for amenorrhoea.

Waltheria indica **L. [Sterculiaceae]**

Karum pootu thalai; erect or procumbent, pubescent herb or undershrub; root used; the root is ground, mixed with honey or milk and given inside for diarrhoea; the root is chewed to control internal haemorrhages; decoction is made from the root and taken in to induce fecundity in women.

Wedelia chinensis **(Osb.) Merr. (*W. calendulaceae* Less.) [Asteraceae]**

Manjal Karisalai; procumbent hairy herb, rooting at nodes; leaves and entire plant are used; the leaf extract is given inside for cough, cephalgia; decoction is made of the entire plant and given inside for menorrhagia and uterine haemorrhages.

Xanthium indicum **Koen. (*X. strumarium* L.) [Asteraceae]**

Aattu otta thalai; unarmed erect herb; leaves used; leaf extract is taken and applied at the effected place for tooth ache, ulcers, boils, abscesses.

Zanthoxylum rhetsa **(Roxb.) DC. (*Z. limonella* (Dennst.) Alst.; *Z. budrunga* (Roxb.) DC) [Rutaceae]**

Palvali pattai; medium sized trees with conical prickles on branches and trunk; bark used; the bark is powdered and applied at the place of tooth ache.

Zanthoxylum tetraspermum **Wt. & Arn. [Rutaceae]**

Milagaranji kirai; large, armed, woody climber; prickles recurred; leaves or root used; the leaves or root are mixed with the leaves of any cacti, *Kleria grandiflora* and wild lime, all ground together, applied all over the body at night and bath taken before dawn to prevent the effects of witchcraft and black magic.

Zehneria maysorensis **(Wt. & Arn.) Arn. var. *maysorensis* (*Melothria mucronata* sensu Chakravar); (*M. perpusilla* var. *subtruncata* Cogn.; *M. perpusilla* sensu Gamble) [Cucurbitaceae]**

Mosu mosukkai; slender climber; leaves used; leaves are boiled with or without the leaves of pumpkin and eaten for good growth of children, fainting attack, head problems, health of pregnant mothers, vermifuge, fever, leucorrhoea, cough and piles.

Zingiber officinale **Roscoe. [Zingiberaceae]**

Inghi (ginger); Slender aromatic rhizomatous herb with leafy stem; rhizome used; rhizome is ground with salt and the extract is given inside. Juice is extracted out of the rhizome and drunk for indigestion, hyperacidity, cough, asthma, dyspnoea, diarrhoea, edema, stomach disorders, anaemia, cardiac diseases, piles, colic, nausea, vomiting, swellon stomach; the extract is added with powdered pepper and given to cattle for any stomach trouble for cattle especially loose motion; dried rhizome is boiled with corriandor and pepper and the extract is drunk for fever; the rhizome is ground with sugar and applied at the affected place for cut injuries; the rhizome and garlic are cut into small pieces, boiled in cow's milk and palmera jaggery is added. This can be preserved for quite sometime. This is eaten daily early in the morning in empty stomach for gas trouble, gastritis; fresh rhizome or dried rhizome is put inside a banana and given to cattle for indigestion and lack of apetite.

Ziziphus oenoplia **(L.) Mill. [Rhamnaceae]**

Chokalam; prickly scandent shrub; used for stall-feeding the goats.

Ziziphus rugosa **Lam. [Rhamnaceae]**

Mullukkottan; straggling armed shrub; fruit, flower and bark used; fruits are collected and eaten for food; decoction is made out of the flowers and given inside for menorrhagia; bark is ground and given inside.

References

1. Aminuddin and R.D. Girach. 1991. Ethnobotanical studies on Bondo tribe of district Koraput (Orissa). India. Ethnobotany. 3: 15-19.

2. Manandhar, N.P. 1996. Ethnobotanical observation of ferns and fern allies of Nepal. J. Econ. Tax. Bot. Addl. Ser. 12: 414-422.

3. Vertak, V.D., and Suryanarayana, M.C. 1995. Enumeration of wild edible plants from Susala island, Mulshi reservoir, Pune district, J. Eco, Tax. Bot. 19(3): 555-569.

4. Kulkarni, D.K. and Kumbhojkar, M.S. 1993. Kitchen garden plants of Mahadeokoli tribe in Maharashtra. Ethnobotany. 5: 119-127.

5. Arora, R.K. and K.R. Mehra. 1977. Special explorations for the collection of plants and their wild relatives in India. Bull. Bot. Surv. India. 19(1-4): 139-144.

6. ESCAP. 1989. Environmental Management of Mountain Ecosystems in Asia and the Pacific. United Nations Economic Commission for Asia and the Pacific, Bangkok, Thailand. ST/ ESCAP/788.

7. Misra, N.M., Mahendra, A.K. and Ansari, M.Y. 1988. Pilot Survey of Fuel Consumption in Rural Areas. The Indian Forester, Vol. 114. No. 2. Feluchar 1988, pp. 75-62.

8. Jenne, H. and Melanie, J. 1989. The Economic Value of Non-Timber Forest Products in Southeast Asia with emphasis on Indonesia, Malaysia, and Thailand. Netherlands Committee for IUCN, Denmark, pp. 175.

9. Falconer, J. 1988. The Importance of non-timber forest products in rural economics: a review of the literature. Report prepared for policy and Planning Services. Forestry Department, FAD, Rome (unpublished manuscript).

10. Food and Agriculture Organization (FAO). 1988. Traditional Food Plants. FAO Food and Nutrition paper No. 42. Rome, pp. 593.

11. Food and Agriculture Organization (FAO). 1989. Forestry and Nutrition: A Reference Manual. Rome, pp. 114.

12. Falconer, J. 1988. The Importance of non-timber forest products in rural economics: a review of the literature. Report prepared for policy and Planning Services. Forestry Department, FAD, Rome (unpublished manuscript).

13. Jenne, H. and Melanie, J. 1989. The Economic Value of Non-Timber Forest Products in Southeast Asia with emphasis on Indonesia, Malaysia, and Thailand. Netherlands Committee for IUCN, Denmark, pp. 175.

14 Hollowas, P.S. and Alexander, G. 1990. Ethnobotany of the Fort Yukon Region. Alaska. Economic Botany, 44(2), April to June, 1990, pp. 214-225.

15. Timbrook, J. 1990. Ethnobotany of Chumash Indians, California. Based on Collections by John P. Harrington.

16. Bhat, R.B., Etejere, E.O. and Oladip, V.T. 1990. Ethnobotonical Studies from Central Nigeria. Economic Botany, 44(3), July-September, 1990.

17. Dennis, P.A. 1988. Herbal medicine among the miskito of Eastern Nicaragua. Economic Botany 42(1), January–March 1988.

18. Manandhar. 1991. Medicinal Plant–Lore of Tumang Tribe of Kabhrepalanchok District, Nepal. Economic Botany, 45(1) January–March 1991, pp. 58-71.

19. Jain S.K. 1981. Ethnobotonical Research Unfolds New Vistas of Traditional Medicine. In: Glimpses of Indian Ethnobotany, (ed.) Jain S.K. Oxford & IBH Publishing Co., New Delhi, pp. 13-36.

20 Singh, H.B. and Arora, R.K. 1978. Wild Edible Plants of India. Indian Council of Agricultural Research. New Delhi, pp. 88.

21. Mittre and Vishnu. 1981. Wild Plants in Indian Folk Life–A Historical Perspective. In: Glimpses of Indian Ethnobotany (ed.). Jain S.K. Oxford and IBH Publishing Co., New Delhi, pp. 37-58.

22. Singh, H.B. and Arora, R.K. 1978. Wild Edible Plants of India. Indian Council of Agricultural Research. New Delhi, pp. 88.

23. Arora, A.K. 1981. Native Food Plants of the Northeastern Tribals, In: Glimpses of Indian Ethnobotany (ed.) Jain, S.K., Oxford and IBH Publishing Co., New Delhi, pp. 91-106.

24. Joseph, J. and Kharkongon, P. 1981. Folklore Medico-Botany of Rural Khasi and Jaintis Tribes in Meghalaya. In: Glimpses of Indian Ethnobotany. Jain S.K. (ed.). Oxford and IBH Publishing Co., New Delhi, pp. 124-136.

Joseph, J. and Khartongon, P. 1981. A Preliminary Ethnobotanical Survey in the Khansi and Jaitia Hills, Maghalaya. In: Glimpses of Indian Ethnobotany. Jain S.K. (ed.). Oxford and IBH Publishing Co., New Delhi, pp. 115-123.

25. Baishya, A.K. and Hajra, P.K. 1981. Ethnobotonical Notes on the Miris (Mistings) of Assam Plains, In: Glimpses of Indian Ethnobotany (ed.) Jain, S.K. Oxford and IBH Publishing Co., New Delhi, pp. 161-169.

26. Borthakur, S.K. 1981. Certain Plants in the Folklore and Folklife of the Karbis (Mikirs) of Assam. In: Glimpses of Indian Enthobotany (ed.) Jain, S.K. Oxford and IBH Publishing Co., New Delhi, pp. 170-181.

27. Vartak, V.D. 1981. Observations on Wild Edible Plants from Hilly Regions of Maharastra and Goa: Resume and Future Prospects. In: Glimpses of Indian Ethnobotany. (ed.) Jain, S.K. Oxford and IBH Publishing Co., New Delhi, pp. 261-271.

28. Abraham, Z. 1981. Ethnobotany of the Todas, the Kotas, and the Irulas of the Nilgris, In: Glimpses of Indian Ethnobotany (ed.). Jain S.K. Oxford and IBH Publishing Co., New Delhi, pp. 308-320.

29. Pangtey, Y.P.S., Samant, S.S. and Rawat, G.S. 1989. Ethnobotonical Notes on the Bhotia Tribes of Kumaun. Indian Journal of Forestry, Vol. 12, No.3. September. 1989, pp. 191-196.

30. Fernandes, W., Menon, G. and Viegas, P. 1988. Forests, Environment and Tribal Economy: Deforestation, Impoverishment, and Marginalisation in Orissa. Indian Social Institute, pp. 363.

31. Fernandes, W. and Menon, G. 1987. Tirbal Women and Forest Economy: Deforestation, Exploitation and Status Change. Tribes of India Series 1. Indian Social Institute, New Delhi. pp. 178.

32. Mohapatra, A.K. 1991. Ethnopharmacological Notes on Pauri. Bhurinya Tribe of Orissa. The Indian Forester, Vol. 117. No. 2. Feb. 1991.

33. Singh, K.K. and Maheswari, J.K. 1990. Plant Wealth in the Life and Economy of the Tharus of Nainital District, Utter Pradesh. The Indian Forester, Vol. 116. No.8. August 1990, pp. 636-641.

34. Maheswari, J.K. 1990. Interaction of Tribals with the Forests. Proceeding of MAB Regional Training Workshop. Tropical Forest Ecosystem Conservation and Development. South and Southeast Asia. (ed.) Nair, K.K.M., Bhat, K.V. and Sharma, J.K. Kerala Forests Research Institution, Peechi, pp. 115-120.

35. Gangwar, A.K. and Ramakrishnan, P.S. 1990. Ethnobiological Notes on Some Tribes of Arunachal Pradesh, Northeastern India. Economic Botany, 44(91) January–March 1990, pp. 94-105.

36. Malhotra, K.C., Poffenberger, M., Bhattacharya, A. and Dev. D. 1992. Rapid Appraisal methodology trials in South West Bengal: Assessing natural forest regeneration patterns and non-wood forest product harvesting practices. Forests, Tress and People News letter, 15/16, p.18-25.

37. Majumdar, A. 1991. Tribal Medicines and their role in community herbal gardens. In: Raychaudhuri, S.P. (ed.) Recent Advances in Medicinal, Aromatic and Spice Crops. 1: 241-247.

Thothathri, K., Sen R. and Pal, D.C. 1989. Status of Medicinal plants in Indian Flora. Journ. Econ. Tax. Bot. 13(3): 605-614.

38. Gupta, T. and Guleira, A. 1982 a. Some Economic and Management Aspects of a Non-wood Forest Product in India: Tendu Leaves. Oxfored and IBH Publishing Co., Pvt. Ltd., New Delhi.

39. Gupta, T. and Guleira, A. 1982 b. Non-wood Forest Products in India. Oxford and IBH Publishing Co., Pvt. Ltd, New Delhi, pp. 1-2.

40. ISST. 1987. Small scale forest enterprises in India, with special reference to role of women: National Review Paper. Institute of Social Studies Trust, New Delhi (mimeo).

41. Chambers, R., Saxena, N.C. and Shah, T. 1989. To the Hands of the Poor, Water and Trees. Oxford and IBH Publishing Co. Pvt. Ltd., New Delhi, pp. 273.

42. Guha, R. 1983. Forestry in British and Post-British India. Economic Political Weekly, October 29, VII. 18, pp. 45-46.

43. GOI. 1982. Report of Committee on Forest and Tribals in India. Ministry of Home Affairs, Government of India, New Delhi, pp. 19.

44. Jenne, H. and Melanie, J. 1989. The Economic Value of Non-Timber Forest Products in Southeast Asia with emphasis on Indonesia, Malaysia, and Thailand. Netherlands Committee for IUCN, Denmark, pp. 175.

45. Sreekumar, K. and Aravindakshan, M. 1984.Creation of a Life supporting system for Tribals–A key Element in the Eco-development of Western Ghats. Paper Presented in the Seminar on Eco-development of Western Ghats,17-18, October 1984.Kerala Forest Research Institute, Peechi, Paper Code 31,pp. 113.

46. Roy Burman, B.K. 1994. Tribes in Perspective. Encyclopaedia of Scheduled Tribes, Scheduled Castes and Disadvantaged Peoples-I, Series-I. MITTAL Publications, New Delhi pp. 343.

47. Chambers, R., Saxena, N.C. and Shah, T. 1989. To the Hands of the Poor, Water and Trees. Oxford and IBH Publishing Co. Pvt. Ltd., New Delhi, pp. 273.

48. Fernandes, W., Menon, G. and Viegas, P. 1988. Forests, Environment and Tribal Economy: Deforestation, Impoverishment, and Marginalisation in Orissa. Indian Social Institute, pp. 363.

49. GOI. 1988. Resolution National Forest Policy, 1988. Ministry of Environment and Forests, No. 3-1/86, FP.

50. Fernandes, W., Menon, G. and Viegas, P. 1988. Forests, Environment and Tribal Economy: Deforestation, Impoverishment, and Marginalisation in Orissa. Indian Social Institute, pp. 363.

51. Lalthara. 1976. Marketing of Agricultural and Forest Produce in Sundaragarh District, (Orissa) with special reference to Lepripara Bloc. The Administrator 21. No.1. Spring 1976, pp. 433-436.

52. Joshi, G. 1983. Forest Policy and Tribal Development: Problems of Implementation, Ecology and Exploitation. In: Towards a New Forest Policy: People's Rights and Environmental Needs (eds.). Fernandes, W. and Kulkarni, S. Indian Social Institute, New Delhi, pp. 25-47.

53. Fernandes, W. and Menon, G. 1988. Deforestation, Forest Dweller Economy and Women. Indian Social Institute, New Delhi.

54. Menon and Geeta. 1987. Tribal Women: Victims of the Development Process. Social Action, Vol. 37.

Menon and Geeta. 1991. Ecological Transitions and the Changing Context of Women's Work in Tribal India. PURUSARTHA, 14. pp. 291-312.

55. Dasgupta and Supriya. 1988. Tribal Women. In: Understanding the Tribal Dilemma–Tribal Women and Forest Dweller Economy. Indian Social Institute, New Delhi.

56. Ali, A. 1980. Health and Nutritional Status of Pauri Bhuniyas of Jaldih village in Sundergarh district, Orissa. The News letter 13 (April), Tribal Development Division, Ministry of New Affairs, New Delhi.

57. Chambers, R., Saxena, N.Ç. and Shah, T. 1989. To the Hands of the Poor, Water and Trees. Oxford and IBH Publishing Co. Pvt. Ltd., New Delhi, pp. 273.

58. Gomp. 1986. Annual Administration Report for 1984-1985. Forest Department. Government of Madhya Pradesh. December 1986. Bhopal (mimeo), pp. 33.

59. GOI. 1988. Resolution National Forest Policy, 1988. Ministry of Environment and Forests, No. 3-1/86, FP.

60. Hollowas, P.S. and Alexander, G. 1990. Ethnobotany of the Fort Yukon Region. Alaska. Economic Botany, 44(2), April to June, 1990, pp. 214-225.

61. Timbrook, J. 1990. Ethnobotany of Chumash Indians, California. Based on Collections by John P. Harrington.

62. Bhat, R.B., Etejere, E.O. and Oladip, V.T. 1990. Ethnobotonical Studies from Central Nigeria. Economic Botany, 44(3), July-September, 1990.

63. Dennis, P.A. 1988. Herbal medicine among the miskito of Eastern Nicaragua. Economic Botany 42(1), January–March 1988.

64. Manandhar. 1991. Medicinal Plant–Lore of Tumang Tribe of Kabhrepalanchok District, Nepal. Economic Botany, 45(1) January–March 1991, pp. 58-71.

65. Arora, A.K. 1981. Native Food Plants of the Northeastern Tribals, In: Glimpses of Indian Ethnobotany (ed.) Jain, S.K., Oxford and IBH Publishing Co., New Delhi, pp. 91-106.

66. Joseph, J. and Kharkongon, P. 1981. Folklore Medico-Botany of Rural Khasi and Jaintis Tribes in Meghalaya. In: Glimpses of Indian Ethnobotany. Jain S.K. (ed.). Oxford and IBH Publishing Co., New Delhi, pp. 124-136.

67. Baishya, A.K. and Hajra, P.K. 1981. Ethnobotonical Notes on the Miris (Mistings) of Assam Plains, In: Glimpses of Indian Ethnobotany (ed.) Jain, S.K. Oxford and IBH Publishing Co., New Delhi, pp. 161-169.

68. Vartak, V.D. 1981. Observations on Wild Edible Plants from Hilly Regions of Maharastra and Goa: Resume and Future Prospects. In: Glimpses of Indian Ethnobotany. (ed.) Jain, S.K. Oxford and IBH Publishing Co., New Delhi, pp. 261-271.

69. Abraham, Z. 1981. Ethnobotany of the Todas, the Kotas, and the Irulas of the Nilgris, In: Glimpses of Indian Ethnobotany (ed.). Jain S.K. Oxford and IBH Publishing Co., New Delhi, pp. 308-320.

70. Ibid.

71. Ibid.

72. Mohapatra, A.K. 1991. Ethnopharmacological Notes on Pauri. Bhurinya Tribe of Orissa. The Indian Forester, Vol. 117. No. 2. Feb. 1991.

73. Singh, K.K. and Maheswari, J.K. 1990. Plant Wealth in the Life and Economy of the Tharus of Nainital District, Utter Pradesh. The Indian Forester, Vol. 116. No.8. August 1990, pp. 636-641.

74. Maheswari, J.K. 1990. Interaction of Tribals with the Forests. Proceeding of MAB Regional Training Workshop. Tropical Forest Ecosystem Conservation and Development. South and Southeast Asia. (ed.) Nair, K.K.M., Bhat, K.V. and Sharma, J.K. Kerala Forests Research Institution, Peechi, pp. 115-120.

75. Gangwar, A.K. and Ramakrishnan, P.S. 1990. Ethnobiological Notes on Some Tribes of Arunachal Pradesh, Northeastern India. Economic Botany, 44(91) January–March 1990, pp. 94-105.

Chapter 8

Alienation and Exploitation of the Tribals

At the National Level

The tribals in India are the aboriginal races-older on Indian soil than the Vedic Aryans and even the Dravidian architects of ancient South Indian civilization. For thousands of years, tribal folk lived undistrubed in hills and densely wooded regions unattractive to the peasantry of more dynamic populations.[1]

Randhawa (1976);[2] Joshi (1979, 1983);[3] Desmond and Alineo (1982);[4] Kannan (1983);[5] Roy Burma (1983);[6] Guha (1983);[7] Royston (1979);[8] Hong and Evelyne (1987);[9] Singh and Chatrapati (1987);[10] Chambers *et al.*, (1989);[11] Poffenberger (1990)[12] and Buddhadeb Chaudhuri (1994)[13] reveal that the forest policies and Acts of the British rulers and later by the Indian Government have progressively alienated the tribals from their traditional habitats and slowly taken away their traditional rights over the forest resources.

Ninety per cent of the tribals in India are below the poverty line. They are entangled in the trap of poverty which extends into all aspects of their lives: susceptibility to disease, limited access to most types of services and information, lack of control over resources, subordination to higher social and economic classes and utter insecurity in the face of changing circumstances. Their poverty trap include skewed patterns of access to land and other assets, physical weakness and powerlessness against corrupt institutions.[14]

While tribals form only 7.5 per cent of the country's population, their proportion among those displaced by development projects is as high as 40 per cent.[15] But fewer than 25 per cent of persons displaced by these projects are rehabilitated properly.[16] To separate the indigenous people from their land is tantamount to tearing them apart from their life-giving source.[17] They understand the individual and the community to belong to the land by virtue of their ancestors being seated in a given territory. The territory is an extension of the Adivasis collective consciousness with a cultural, political and social significance[18]. Tribal loss of land is a loss of human dignity, culture, community and the community resource base.[19]

Having first been driven over centuries to retreat into refuge zones such as hills and forests, the tribals are now being forcibly pushed out of an ambience with which they have gradually developed a close relationship.[20] Invariably, there is degeneration in their lives and they are forced to become migrant labourers or urban slum dwellers.[21]

In some places, they organise, resist and assert their rights. Thus, whether in Chhattisgarh, Jharkhand, Orissa, Maharashtra, Madhya Pradesh, Tamil Nadu, Karnataka or Kerala, the struggle of tribal communities for their rights is inextricably linked to land and forests.[22]

Tribals have lived and cared for forest areas for centuries. Of late, their presence in forests has been challenged, their lifestyle misunderstood and their existence criminalised so as to label them trespassers, guilty of encroachment. They have no political edge and are defenceless when an administration decides to move against them. Tribals are being thrown out of forests. They are being harassed and beaten.[23] Forest to tribals means something more than a source of forest produces. Their habitat, economic life and society have incredible linkage with the forest.[24]

Geographical fragmentation of the Adivasis ensures that they remain politically isolated, suppressed or insignificant in most places. Such divisions, along with modernization and homogenization have suppressed their languages and cultures, pushing them into the cultural mainstream with dire consequences. A systematic internal colonization in some places has brought Adivasi communities to the brink of ethnocide. The new economic policies, consequent to the adoption of globalization and liberalization, have further intensified the threats to the survival of Adivasis.

To add to the woes of the Adivasi community, the officials indulge in large-scale embezzlement of funds meant for their development. Redress in cases of official and non-official harassment is denied to the Adivasis and they are unable to cope with the mysteries of a multi-tiered bureaucratic apparatus.[25]

In the Palni Hills

Expansion of human habitats and commercialization of Palni hills resulted in drastic depletion of forests which were a source of livelihood for these two tribal groups. Down the centuries, various other groups inhabited the hills and deprived the tribals of their lands, forests and habitats. The Kunnavans (Vellalans) from the Coimbatore plateau came in the 14th or 15th centuries probably to escape the terror of war or famine. They settled chiefly in the lower Palnis, dispossessing the tribals and energetically farming their lands.[26] In the 18th century came the Telungu speaking 'Chettiars', followed by foreigners, missionaries, and British officials. Twentieth century saw a tremendous influx of population and expansion of human settlements and estates. Today, the aboriginals are serfs and coolies and considered as outcasts.[27] These tribals are at present under extreme pressure from economically exploitative and socially dominant Tamils from the plains.[28]

The basic features of the changing situation are four in number:

1. Under great pressure for the resources of the forests, Tamil plainsmen are virtually flooding into the tribal habitat
2. Transformation of forests into plantations
3. Introduction of benevolent agents, like social action groups and Government agencies

4. The lure of material advantages (movies, plastic combs, metal utensils, quality tools and clothing) and thus a sharp break from the past.

Deprived of their lands and denied of their rights, the Palliyans and Pulayans in the Palni Hills now lead a life of misery in the 'civilised' jungles of coffee estates and other plantations in the hilly terrains of Kodaikanal. With 'civilisation' entering the jungles, with avaricious men taking over their lands by unfair means, exploiting their innocence, they had been driven out of their traditional habitations. According to Palni Hills Adivasis Liberation Movement general secretary Leelawathi, the Central and State governments allocate crores of rupees every year for the welfare and upliftment of the tribals. But whether even a minor percentage of this amount is really spent on the welfare of the tribals is doubtful, taking into consideration, the appalling conditions under which they live, alienated from their traditional lands.[29]

They work as bonded labourers in their own lands, usurped from them by the men from the plains. Theirs' is a tale of tragedy and trauma from cradle to grave. Pregnant women are usually carried in hammocks for four or five km. to reach the main road and then take a bus to the nearest hospital. When there is a death in their families, they have no burial grounds to bury the bodies. These were tear jerking stories related by tribal women from the villages, scattered among the leafy gray atmosphere of Kodaikanal hills in Dindigul district at the public hearing on 'Problems facing the Adivasi and Dalit women in Kodaikanal' held on 19th September 2002. No potable water was available in many villages and they had to walk miles to fetch water. Thus, 'the sons and daughters of the soil' have been forgotten by time and development.[30]

Disappearing Traditional Lifestyle

Significant elements of their traditional lifestyle and value system include:

1. Community living and common ownership of resources: Earlier they were nomadic tribes and moved from place to place in groups in the forests.
2. Their culture and value-systems were transmitted to the new generation through the medium of dances, songs, myths and folklore.

3. They were very much conscious of natural justice. They upheld values like truth, generosity, genuinness, simplicity, neighbourly care and cosmic union (treating the animals and trees as part of their race).
4. Deep consciousness of their relatedness to and dependence on their ancestors, others and nature.
5. Equality of sexes and respect for elders.
6. After the invasion of the 'Gaunders', Kunnavans, Chettiars, Britishers and others from the plains in the 18th to 19th centuries, the forest resources have been ruthlessly exploited and these tribal communities have been more and more alienated from their original habitat and consequently exploited socially and economically. Living outside the forests, most of them have been reduced to be daily-wage earners in the coffee estates and they are paid very meager wages.

Since they are now settled outside the forests they no more depend on the forests completely for their subsistence and other needs and they no longer view the forests as their 'home'.

Their traditional lifestyle and value-system have been changing much due to the following factors:

1. Loss of their traditional habitat–the forests.
2. Economic dependence on and close association with the estate owners and the resultant influence of the dominant culture and dominating values of the owners and employees on the tribals.
3. Influence of mass-media, especially television, cinema and radio.

The notable changes in the tribal lives in the Palni hills are:

1. *Change in the lifestyle*: their self-sufficient economy has been replaced by market economy and from hunters and gatherers, they have become daily-wage earners. From nomadic life, they have a settled way of life now. They no longer move from place-to-place.
2. *Food habits*: from yams, greens and honey to rice and curry and the consequent health problems due to their lack of

nutritional content in their food intake. Still births and infant mortality are on the increase among them. Life-expectancy is gradually declining.

3. *Dress and ornaments*: from grass and leaves to modern dresses and from natural ornaments to artificially made ones.
4. *Recreation*: from dance and music to radio, films and video and their influence on the thinking, behaviour and culture.
5. *Health care*: from herbal medicines to allopathic medicines. Most of them have completely forgotten their traditional wisdom about herbal medicines.
6. *Attitude towards forests*: outside dependence for all their needs has resulted in their callous and even destructive attitude towards the remaining forests. The whole forests were theirs once and rampant deforestation and expansion of estates have introduced the desire of 'ownership' into their culture. The tribals themselves have cleared/are clearing the forests as they want to 'own' some piece of land for cultivation.
7. *Changes in their religious practices*: Previously they celebrated the festival of their deities only once in a year. Now they have started celebrating popular festivals like Deepavali, Pongal, Karthigai etc. and popular devotions like fasting and pilgrimage to Palni Murugan temple and Sabarimalai Iyyappan temple. There were no structures earlier for their places of worship. Generally their deities were kept in the forests. But now they have started putting up some structures as their temple and that too in their own hamlets. Thus they have started accepting and practicing some alien religions and religious practices. In places like Moolaiyar, the estate-owners themselves have helped the tribals economically to build temples for alien gods and goddesses.
8. These tribals have now gone in for nuclear families from group living. They limit their relationship now within their families and within their kith and kin.
9. Arranged marriages are taking the place of free marriages gradually.

10. There is a growing male domination and chauvinism.
11. Market economy has completely ruined their life. They are not able to cope with the sudden onslaught of this new situation. This again has increased their dependence and slavery to their employers.
12. From their original life of freedom, they are now very much oppressed economically and culturally by their employers. Women are exploited and abused sexually by the estate owners. Now there are even cases of S.T.D among them.

References

1. Christoph Von Fiirer–Haimendorf. 1985. Tribes of India. The struggle for survival. Oxford University Press, Delhi p. 342.
2. Randhawa, R.K. 1976. The Forest in the Life of the Tribals. The Administor, 21, b. 1. Spring 1976, pp. 76.
3. Joshi, G. 1983. Forest Policy and Tribal Development: Problems of Implementation, Ecology and Exploitation. In: Towards a New Forest Policy: People's Rights and Environmental Needs (eds.). Fernandes, W. and Kulkarni, S. Indian Social Institute, New Delhi, pp. 25-47.

   Joshi, G. 1983. Forests and Forest Policy in India. Social Scientist 11, No.1, January 1983, pp. 43-52.
4. Desmond and Alineo, D. 1982. People and Forests: The Forest Bill and a New Forest Policy. Monograph Series, (general ed.) Alfred de Souza. Indian Social Institute, New Delhi, pp. 49.
5. Kannan, K.P. 1983. Forestry Legislation in India, Its Evolution in the Light of the Forest Bill. 1980. In: Towards a New Forest Policy: Peoples Rights and Environmental Needs (eds.). Fernandes, W. and Kulkarni, S. Indian Social Institute, New Delhi, pp. 75-83.
6. Roy Burman, B.K. 1983. 'Issues for a forest policy', in Mainstream, October 22, 1983. p. 127.
7. Guha, R. 1983. Forestry in British and Post-British India. Economic Political Weekly, October 29, VII. 18, pp. 45-46.
8. Royston, M.G. 1979. Development Benefits of Environmental Management Policies. In: Ecology and Development (eds.)

Bandhu, D. and Bhardwaj, W. Indian Environmental Society, New Delhi, pp. 7-32.

9. Hong and Evelyne. 1987. Natives of Sarawak, Survival in Borneo's Vanishing Forests. Institute Masyarakat, Malaysia, pp. 259.
10. Singh and Chatrapati. 1987. Forest and the Law in India. The Indian Law Institute, New Delhi.
11. Chambers, R., Saxena, N.C. and Shah, T. 1989. To the Hands of the Poor, Water and Trees. Oxford and IBH Publishing Co. Pvt. Ltd., New Delhi, pp. 273.
12. Poffenberger and Mark. 1990. Joint Management of Forest Lands: Experiences From South Asia. A Ford Foundation Programme Statement, pp. 49.
13. Buddhadeb Chaudhuri. 1994. 'Social and Environmental Dimensions of Tribal Health'. In: Basu, Salil (ed.). Tribal Health in India. Manak Publications Pvt. Ltd. p. 81.
14. Durning, A.B. 1990. Ending Poverty. In: State of the World 1990. (ed.) Starke. L.A. World watch Institute Report on Progress Towards a Sustainable Society. W.W. Norton and Company, New York, London, pp. 135-153.
15. GOI. 1985. Report of the Committee on Rehabilitation of Displaced Tribals due to Development Projects. Ministry of Home Affairs, New Delhi, pp. 18-19.
16. Hansda and Subotu. 1983. Agricultural Development in Tribal Areas. In: Mishra and Bhapnder (eds.). Tribal Area Development Society for the Study of Regional Disparities, New Delhi, pp. 21-26.
17. Fernandes, W. and Thukral, *E.G.* 1989. Introduction: Questions on Developments Displacement and Rehabilitation. In: Development Displacement and Rehabilitation, (eds.) Fernandes, W. and Thukral, G. Indian Social Institute. New Delhi, pp. 1-10.
18. Bijoy, C.R., Prabhu, P. 1997. Adivasi Situational Status,–III. International Alliance of Indigenous–Tribal Peoples of the Tropical Forests, Nagpur, India, 3-8 March p.7.

19. Singh and Chatrapati. 1989. Rehabilitation and Right to Property. In: Development Displacement and Rehabilitation (eds.) Fernandes, W. and Thukral, *E.G.* Indian Social Institute, Delhi, pp. 91-103.

20. Mihir Shah. 2002. 'The Adivasi question–I, in 'The Hindu', April 2, 2002. p. 10.

Mihir Shah. 2002. 'The Adivasi question–II, in 'The Hindu', April 3, 2002. p. 10.

21. Fernandes, W. and Thukral, E.G. 1989. Introduction: Questions on Developments Displacement and Rehabilitation. In: Development Displacement and Rehabilitation, (eds.) Fernandes, W. and Thukral, G. Indian Social Institute. New Delhi, pp. 1-10.

22. Kalpana Sharma, 2003. 'Adivasis–the forgotten India', in 'The Hindu', Sunday, March 9, 2003 p. 16-17.

23. Rajeev Dhavan, 2002. 'Evicting 10 million tribals' in 'The Hindu', September 20, 2002, p. 10.

24. Anil Kumar Singh (ed.) 1993. Tribes and Tribal Life–Tribal Culture and Policy–Vol. 2. SARUP and Sons. p. 389.

25. Kuldip Nayar. 2002. 'Encroaching on tribal land' in 'The Hindu'. February 27. 2002. p.10.

26. Mitchell and Nota. 1972. The Indian Hill-Station: Kodaikanal. Department of Geography. Research paper No. 141. University of Chigago, pp. 191.

27. Wyckoff, C.C. 1984. Kodaikanal–Gift of the Forest. The Christian Literature Society, Madras.

28. Gardner, P.M. 1965a. Ecology and Social Structure in Refugee Populations: The Paliyans of South India. Unpublished Ph.D. Dissertation. University of Pennsylvania, Philadelphia, Pennsylvania, pp. 76-107.

Gardner, P.M. 1965b. Paliyan Social Structure. Paper Presented at the National Museum of Canada Conference on Band Organization. National Museum of Canada Bulletin, pp. 90-105.

29. Express News Service 2002. Palni Hills Adivasis go in search of their duties. March 26.

30. Krishnaswamy, P. 2002. 'Distressed, displaced tribal women of Kodai hills relate tales of woe', in 'The New Indian Express', September 21, 2002. P.4.

Chapter 9

Tribal Development and Conservation of Forests

Some Conclusions from the Study

As against the data presented and by the impassionate assessment of their socio-economic status, the following conclusions can be drawn:

1. The Palliyans and the Pulayans are the earliest known occupants of these hills within historic times
2. The tribal villages are predominantly spread over the Lower Palnis
3. Both Palliyans and Pulayans have a lot of physical and cultural similarities
4. Most of them do not own either houses or cultivable lands
5. Most of them are daily wage earners and they work in the coffee estates and are paid very low wages

6. Collecting non-timber forest products remains the principal subsidiary occupation
7. They are educationally backward
8. Tribal customs and practices are still prevalent among them in spite of the onslaught of alien culture and its influence on them
9. With regard to health care, there are two types of healers among them and they ritualists and herbalists
10. Existing interaction with the remaining forests:
    (*a*) They collect 326 plants from the forests for medicine, domestic use, food, culture and commercial purposes. Among these plants, 233 are found to be medicinally useful. They are used to cure 192 diseases.
    (*b*) Most of these medicinal plants are roadside weeds and pathway plants in the forests and so they are easily available and accessible
    (*c*) Collection of these plants is only need based and no commercial activity is involved in this
    (*d*) Collection of NTFPs does not impringe on the survival of these species. The tribals do not overexploit these resources.

Tribal Health Care

It is evident that the herbal medicine is the first level of contact to tribals when they require medicare. The tradition of tribal medicine evolved from time immemorial, as an outcome of the direct interaction of tribals with their natural environment. The tribal system of traditional medical wisdom is made up of beliefs, norms and practices that have been passed down through an oral tradition. Thus it expresses itself in the empirical stream and is practised by the innumerable herbal medicine healers in tribal communities, who treat a variety of common and chronic ailments. Tribal medicine represents a highly decentralised knowledge of health care that is indigenous, local resource-based, community-rooted and region-specific. The regional bio-diversity–trees, shrubs, grasses, ferns, orchids, climbers, runners, tubers-forms the major resource base of these traditions. Over 6,000 species of plants are used in these traditions, along with the more limited use of animal products and minerals, all over the subcontinent from Ladakh in the north to Kanyakumari in the south. India has one of the richest ethno-botanical traditions in the world. In fact, local health traditions are found all over rural south-east Asia, but they have gone unnoticed because of the western ethnocentric views of health policymakers. So it is imperative for any Government to take immediate steps to introduce the use of herbal medicines to supplement primary health centres. Health training and health awareness programmes can incorporate this knowledge about indigenous medicinal systems.

There is a need to explore further the medicinal properties of plants, which are readily available and extracts of animal and mineral substances used in folklore medicines, through careful observation and validation for application.

There is a need to establish financial support to promote the potential folklore medicines in primary health care. Inventory and documentation of various medicinal plants and herbs, which are in use in the local area, should be developed. For the preservation of various medicinal plants, establishment of botanical gardens, community and kitchen gardens are necessary. This ensures a sustainable supply of safe, effective and affordable medicinal herbs. This type of initiative will enable the developing countries to look inward rather than continuing to rely on expensive, imported medicines having side effects.

A good networking between all research centres and researches is required so that information is collected and there is no duplication of work and waste of resources. The government agencies and financial institutions *viz.* MOEn, DST, DOEn, etc. could be contacted exploring funds in formulating proposals on medicinal plants cultivation and judicial collection of herbs from natural resources.

With the legacy that we have inherited a rich biodiversity and an enormous wealth of knowledge, India should be a pioneer in effecting an amalgamation of whole and standardised extracts of plants into an essential list of herbal medicines for the country and the world.

Modern Medicine Versus Traditional Medicine

Some Drawbacks of Ethnomedicine

The knowledge about ethnomedicine is a part of the traditional lore of the community. This secret knowledge about medicinal plants and magical rituals are always transferred orally through generations and there are no written documents to store this knowledge.

The ethnomedical system of the present age is in a more or less static condition. Research and development are concepts that are unknown to the tribal medicinemen. They follow the age-old medicinal combination and healing techniques that they have inherited from their forefathers. This medical system is a closed one and it is neither experimental nor does it imbibe any knowledge from other cultures in any systematic way.

Ethnomedicine often cannot cope with the fast changes occurring in the community. Changes occurred in the habitat, subsistence pattern and dietary habits of the tribals and the close association with non-tribals tilted the balance of their traditional life. Along with these changes, acculturation brought forth many new diseases to them. Most of these diseases were not present among them in the past and hence the tribal medicinemen are helpless to tackle them. It is quite natural that there is no medicine for maladies like pollution and adulteration in nature. The medicinemen are not able to diagnose these causes which are not familiar to them.

The medicinemen of the present age are not so competent as their predecessors. As the tribals in the past solely depended upon the forest for their food and medicine, the medicinemen of those ages had a vast knowledge of the medicinal properties of flora and fauna. The tribals gradually lost their close familiarity with the forest, when they changed their subsistence pattern from shifting cultivation to settled agriculture. With the death of the old medicinemen, considerable amount of knowledge was lost to the community. There are also some professional hindrances that retarded the passing of medical knowledge from a medicineman to his disciples. It is a commonly accepted notion that no teacher must teach all his techniques to his disciples. So the teacher used to hold back some of the important techniques from the disciples in order to keep his supremacy. Some of the teachers want to reveal this knowledge in their last days. But very often they will not get any opportunity to do so and the precious medical knowledge will be lost forever.

Deforestation plays a role disadvantageous to the ethnomedical system. The large-scale destruction of natural forest adversely affected both flora and fauna of this part of the country. A number of medicinal plants have been destroyed by the ruthless denudation of forest and today the medicinemen suffer from the scarcity of these plants. As the tribals do not have the habit of growing medicinal plants for their needs, they have always depended upon natural sources. When these natural sources are destroyed they find it difficult to collect the various essential ingredients of medicinal preparations. This is equally applicable in the case of animals also. They used to mix a number of animal matter as ingredients of medicines. But deforestation affected their availability also.

The expertise of tribal medicinemen in the art of surgery is very crude and not sufficient for the successful treatment of complicated cases. Even though they have a good knowledge in the bone-setting technique, they may commit mistakes when fractures are involved. Due to the mistakes in the treatment, some of the patients are found handicapped.

The skeptical attitude of the patients is a new problem confronting the practitioners of ethnomedicine. After getting little education and experiencing the capacity of modern medicine to impart speedy relief, the tribals, especially the youngsters are more skeptical towards the capability of ethonomedical practices than elders. This disbelief often retards the pace of healing process. When the cure comes very slowly, the patients may go to a hospital for availing the facilities of modern medicine for a speedy relief.

Strengths of Ethnomedicine

In spite of the aforementioned drawbacks, ethnomedicine has the following positive aspects:

1. Easy availability
2. Economy–available free of cost
3. Efficacy–rituals associated also has a psychological healing effect

4. Purity–as plant, mineral and animal materials obtained from nature are used for the preparation of medicine.
5. Healer–patient relationship: the practitioner of ethnomedicine is an insider to the community and the patients have personal relationships with him. The healer is related to the patient as a friend, kinsman etc. So their relationship is very cardinal and informal.
6. No side effects
7. Area / region specific

The tribal people can be advised to retain the positive aspects of their ethnomedicine. This traditional mode of health care proved its versatility and efficiency by preserving the health of people for centuries. While introducing the modern ways of health care, special efforts must be made to enlighten them about the utility of their ethnomedicine. The knowledge of the tribals about their indigenous medicines can play an important role in strengthening modern primary health care. They are economical and culturally accepted programmes of community health care.

The ethnomedicine of tribals is in a process of change due to the interaction with other cultures. Initially, the non-tribal immigrants from the plains influenced the tribal culture in a big way. Afterwards, the introduction of modern medicine into the tribal universe by both governmental and private agencies affected ethnomedicine and brought changes into this system. However, the nature of interaction between modern and ethnomedicine and the subsequent social change among various tribal communities are not similar everywhere.

Even in this transitional period, considerable number of ethnomedical beliefs and practices are still persisting in the tribal communities. It is evident that, in the concepts of etiology, in the treatment of diseases and in the prophylactic measures, the traditional medical beliefs and practices are still prevailing. It is also noticed that the various elements of ethnomedicine are reinforced by the principles of tribal religion.

Tribals take to modern medicine only when the illness is an imported one or one which cannot be explained in terms of tribal logic. It is understood that the tribals possess effective methods in their ethnomedicine to treat the diseases which are present among them since the remote past. However, they do not know how to heal the newly imported diseases and, therefore, seek the assistance of modern medicine.

The global market for herbal healthcare is estimated to be around 62 billion, while India roughly accounts for Rs.2,500 to 3,000 crores (herbal as well as personal care).[1] Officials in the Institute of Indian medicine and Homeopathy say Tamil Nadu has inherent strengths including conducive agro- climate zones to tap the potential of the Fast Moving Herbal Goods (FMHG) market. Moreover, the country has 8,000 of the 21,000 species (medicinal plants) identified by the World Health Organization.[2] Treatment using herbal medicine may be slow but definitely has a long-term effect.[3]

Utilisation of the traditional systems of medicine for promotive, preventive and curative health services is being seriously considered in many parts of the world.[4] The Government should play an active role in promoting this. This will pave the way for tribal development too.

Prospects of Revival of Ethnomedicine

Attempts to revitalise folk medicine must start at the policy level itself. Nurseries nurturing plants of medicinal value ought to be encouraged in every ecosystem. Booklets containing information on local medicines should be popularised.

The whole system of tribal medicinal practice needs to be nurtured carefully. Healers have always perceived their work as a service to the needy, never as a commercial proposition. They also enjoy full community support. If the government health departments were to co-opt their services for a salary, this would spell disaster as it would cripple the autonomy of the tribal traditions.

It is in this context that the Lok Swasthya Parampara Samvardhan Samithi (LSPSS) has been established. One of its activities is focussed on conducting training programmes for PHC (Primary Health Centre) workers. They are taught to identify the local medicinal plants and make simple medicinal preparations that have been traditionally used by the local people to combat common ailments. The Samithi functions by keeping in touch with the folk and tribal practitioners and gaining information about their methods and practices. It also promotes the growth of nurseries serving as resource-bases for traditional medicinal plants.

Tribals and other communities that are dependant on their ecosystems discovered the bounty of nature and developed traditions of health care, food and nutrition, housing and agriculture-based on local resources. Revitalising these and other indigenous traditions is the appropriate way to achieve self-reliance. A revitalisation programme has three practical elements.

1. To build on what is sound and relevant
2. To add to what is incomplete
3. To discourage all that is distorted.

A culturally sensitive development is the natural way to modernise. Tribal and other folk systems of medicine, along with classical systems of traditional medicine, can help India grow more self-reliant and modernise its public health.

Media and communication could play a major role in conservation efforts. They could be used effectively to disseminate different messages for a varied audience to bring about a change in the existing public notions and perceptions that whatever is old and traditional is to be discarded.

The inadequacy of the 'English medicines' is now getting increasingly realized due to its high cost and side-effects, overcrowding of city hospitals and non-availability of the modern doctors in the villages. Coupled with this awareness is the enormous resurgences of interest in herbal remedies and traditional knowledge systems and practices. The importance of traditional medicines to provide low-cost and safe health care to a vast majority of people who are otherwise poor is to be seriously considered. Unfortunately, due to lack of support and recognition to traditional medicine, enough harm has been already done. Enough publicity is already generated on the conservation and sustainable utilization of medicinal plants and mutual sharing of benefits by the Convention on Biological Diversity and the GATT treaty. The only option available at present seems to be to augment medicinal plant resource base and production of genuine herbal drugs affordable to common man through encouraging medicinal plant cultivation and treat important medicinal herbs on par with conventional crops. Medicinal plants cultivation is a need of the hour and should be achieved at all cost to ensure 'health far all in the new millennium'.

Conservation of the Medicinal Plants

Conservation, the main concern of the times, can be effectively done by creating ethno-medicinal parks. An ethnomedicinal park is a collection of native medicinal flora managed with the involvement of ethnic communities. One such park has come up in India in a district of Karnataka, Tumkur, known for its human-made reservoirs for rainwater. The park has a surface of 30 acres located in a village called S. Lekkehalli, 10 kms. from Tiptur town of Tumkur district and was started in 1997. It was conceived with an idea of enhancing appreciation of the local cultures and its knowledge, as plants are a vital link between biodiversity and cultural diversity. The park is home to 324 species of native medicinal plants, 101 trees, 49 shrubs, 46 climbers and 128 herbs. This diversity represents nearly half the total number of medicinal plant species of 686 in the district.[5] In this connection, the following measures can be undertaken in the Palni hills:

1. Conservation of the habitats of these important plants.
2. Systematic surveys, status assessment, habitat management restoration.
3. In order to prioretise species, one needs a good idea of their status, distribution, rate of growth or decline, threats etc.
4. For effective conservation of genetic resources of medicinal plants, we urgently need a complete check- list of threatened plants region-wise.
5. To provide suitable protection to the threatened medicinal plants, nature reserves can be established. Conservation through cultivation should be encouraged.
6. Establishment of 'germplasm bank' and 'seed bank' is very essential for medium and long-term storage of seeds, pollen and clonal materials.
7. Medicinal plants gardens or farms should be developed for *'in situ'* and *'ex situ'* conservation. These gardens / farms should preferably be managed by the local tribal communities who can reintroduce, preserve and cultivate the local genotype of medicinal plant species.

One such ethnomedicinal park can be established in Perumalmalai (about 12 kms. down Kodaikanal town). There is already a Botanical garden, established and maintained by St. Xavier's College, Palayamkottai. Part of the garden can be an ethnomedicinal park. As there are a few tribal villages around this place, the tribals can be involved in the establishment and maintenance of this park.

Tribal Development

In June 1957, an International Convention (Convention No. 107) in respect of indigenous and other tribal and semi-tribal population was held. This convention laid down that the member states should enable such populations to pursue their material well being and spiritual development in conditions of freedom and dignity, of economic security and cultural specificities which hinder them from benefitting fully from the rights and advantages engaged by other elements of the population and also help the process of integration of such populations into the national communities.[6]

Regional and global meetings on the rights of indigenous peoples are now commonplace. In June 1992, three separate conferences of indigenous peoples were held in Rio de Janerio in connection with the UN conference on Environment and Development.

Principle 22 of the Rio Declaration states, "Indigenous people and their communities and other local communities have a vital role in environmental management and development because of their knowledge and traditional practices. States should recognise and duly support their identity, culture and interests and enable their effective participation in the achievement of sustainable development".

For tribal people the key to their survival is recognition of their rights, their rights to the inalienable collective ownership of their traditional territories; their right to an acceptance of the legitimacy of their own institutions and systems of decision-making.

Most of the countries in the South are developing countries. They are fortunately in the tropics, rich in biodiversity. But the temperate developed North wants to have a free and full access to the biological wealth by declaring these as common property resources for the entire world. Tribal involvement is pivotal in the guard against this piracy of biodiversity. Our biological wealth can be actively preserved by protecting the inalienable rights of the tribals over these resources.

To restore the traditional livelihood of the tribals, there is a dire need of new Forest Policy which should go beyond the colonial model (which views the Forests primarily as a source of revenue and industrial raw materials) and involve forest dwellers who also can have a permanent vested interest in forest conservation.[7] Protection of forests can only be ensured with the willing participation of the population whose life is dependent on forestry.

Towards a New Forest Policy

Any new forest policy should take into account the following guidelines:

1. Any policy should look upon the people as the rightful guardians of the forests and the management of this common property resource should be by the people themselves through decentralized community based institutions.[8]
2. The livelihood rights of the poor people must be paramount and the main thrusts of policy should be based on the feasible rather than with the ideal and starts with organizations and people where they are, with what they control and with their comparative advantages in access and capability[9]
3. Forest dwellers should be reorganised to have a right to be consulted in the making of laws that affect them and any decision taken must be effectively communicated to the people.[10]

Thus any new forest policy should be geared towards meeting the basic requirements of fuel, fodder, food and construction material of the tribal and rural people and it should generate employment for the rural poor and enhance their standard of living.[11]

Constitutional Safeguards for the Tribal Development

Under the broad matrix of plans, a number of programmes concerning education, economic development, rehabilitation, health and nutrition, drinking water facilities, protective measures for the interests and the cultures of tribals have been launched by the government of India.[12]

Article 46 which is otherwise known as the Directive Principles of the State Policy lays down that the State shall promote with special care the educational and economic interest of the weaker sections and in particular of the Scheduled Castes and Scheduled Tribes. The State shall also protect them from social injustice and all forms of exploitation. Article 244 empowers the President of India to declare an area as a Scheduled or Tribal Area under the Fifth Schedule and/ or the Sixth Schedule separately.

The constitution of India provides a number of safeguards for the Scheduled Castes and Scheduled Tribes. All these safeguards have apparently been provided to facilitate the implementation of the Directive Principles contained in article 46 of the constitution.

Article 339 (2) lays down that the executive powers of the union be extended to the giving of directions to a State as to the drawing up and execution of schemes for welfare of the Scheduled Tribes in the state.

Article 275 provides for financial assistance to the State for implementation of schemes of development of Scheduled Tribes and raising the level of administration in the Scheduled areas.[13]

In July 1958, our late Prime Minister, Jawaharlal Nehru explicitly enunciated the strategies for tribal development in more clear terms summarising them in five fundamental principles which are commonly known as 'panchasheel' for tribal development. They are:

1. People should develop along the lines of their genius and we should avoid imposing anything on them. We should try to encourage in every way their own traditional arts and culture.
2. Tribal rights to land and forests should be respected.
3. We should try to train and build up a team of their own people to do the work of administration and development. Some technical personnel from outside will, no doubt, be needed especially in the beginning. But we should avoid too many outsiders into tribal territory.

4. We should not over administer these areas or overwhelm them with a multiplicity of schemes. We should rather work through and not in rivalry to their own social and cultural institutions.
5. We should judge the results not by statistics or the amount of money spent but by the quality of human character that is evolved.

A separate department for the Adi-Dravidar (Scheduled Castes) and the Tribal welfare was created in 1949. The main objective of the department is to implement several development and welfare programmes of the SCs and STs, keeping in view the commitment of the Nation for the promotion of educational and economic interests of these classes of society, enshrined in the Directive Principles of the State Policy.

Tribals in Tamil Nadu are covered under three categories:

1. Tribal sub plan: Certain clusters of tribal pockets are treated under this. This is operative in Salem, South Arcot, North Arcot, Thiruvannamalai, Dharmapuri and Trichy districts. About 1.88 lakhs tribals are covered in this.
2. Primitive tribals: Three characteristics make a tribe primitive one:
   (*a*) Dwindling population
   (*b*) Low level of literacy
   (*c*) Pre-agricultural level of technology.
   About 1.44 lakhs tribals are covered in this in Tamil Nadu.
3. Displaced tribals: All the other tribals who are not covered by the above two categories are included in this. About 2 lakhs tribals are covered in this.

Development of the Tribals in the Palni Hills

In keeping with the aforementioned recommendations and 'keeping in mind the need of these tribals for a livelihood, the greed of middlemen for quick profits, the demands of nature and short-sightedness of the administration' (Smt. Indra Gandhi, 1982–Report of the Task Force for the Study of the Eco-development in the Himalayan Region), the following measures are suggested for the welfare and upliftment of the tribals in the Palni hills.

Immediate measures should be taken at protecting the tribals against unscrupulous landlords, lease-holders and curious money lenders who exploit these poor people as 'working machines' and push them around as 'old style slaves'.[14]

There are a lot of NGO groups like Women's Unity Forum of Tamil Nadu, KEDS, MOTHERS, KALANGIAM, SPIRIT, KARUNYA, VAHRID, SIDT etc. working among these tribal people. How much of development they have brought into the lives of these people is a big question. One is tempted to raise the following questions while analysing the approach and work of these action groups.

1. Do these groups give utmost priority to the people and their needs or they revolve only around their approved projects and funds? Are the tribals fund-raising and profit-making guineapigs?
2. These NGOs have been working atleast for the past one decade. In spite of this, there is no sufficient awareness, development and unity among these tribals.
3. One wonders whether these action groups have made these tribals very much 'dependent' on them. How far these groups have helped in asserting and establishing the tribals' individuality, uniqueness, human dignity and tribal identity is highly doubtful.
4. Some NGOs, though they say, they are committed to the people, have wound up their programmes.
   (*a*) Is it because their projects and funds have been exhausted?
   (*b*) Lack of committed workers!
   (*c*) Wrong and inappropriate approach and plans!

It is found that many NGOs working in the same tribal village with their own aims and projects confuse the people. Rivalry among these NGOs also alienate the people and cause division among them and polarise their loyalties.

One appreciable attempt at educating the tribal children is being carried out by PEAK (Peoples' Education and Action in Kodaikanal hills), an NGO team attached to Sacred Heart College at Shembaganur. So far they have educated 49 tribal students and are educating 20 tribal students now. Their attempt bears good result, because of their positive approach and good rapport with the tribals for the past two decades. Formal education is imparted to the students through play-way method. Such efforts are to be recognised and encouraged.

Though the tribals in the Palni hills put in a lot of labour in NTFP collection, the price they receive for each product is drastically lower than the selling price in the market. The middlemen who take the greatest share of the profit should be completely eliminated. The only way to do this is by forming a tribal co-operative for the NTFPs. The tribals are traditionally skilled[15] in the collection of forest products. By forming a co-operative and by giving a better return for the labour input in the collection, more and more people will be persuaded to take up this one as their regular employment. No more dependence on the estates and estate-owners is necessary if round the year employment is available in NTFP sector. Thus all the economical (paying lower wages demanding more work) and sexual (making use of the tribal girls and women who go for work) exploitations of the tribals by the estate-owners can be averted.

Two village level meetings were organised for the tribals in Varagaraparai during the present study. In both the meetings, the tribals favoured, in fact, warmly welcomed the idea of co-operatives for NTFPs. All of them realize that they are paid low price for the NTFPs collected but they feel that there are no other alternative to this. However, they feel that with the proposal of co-operatives, a new ray of hope has dawned on them.

The tribals can be collectively involved in the management of the forest resources in the Palni hills. Some examples can be quoted for common property resources management from other places in India.

Common Property Resources and People's Participation

The FAO (1982)[16] defines people's participation as "the process by which the rural poor are able to organize themselves and through their own organizations, are able to identify their own needs, share in the design, implementation and evaluation of the participatory action".

The only way to achieve equity and sustainability is by deepening democracy and participation at the village level as much as possible. Every settlement in the country must have a clearly defined environment to protect, care for and use and an open forum in which all can get together to discuss their problems and find common solution.[17] Recent experiences in joint management indicate that when communities are empowered and successfully organized to protect degraded resources, ecological regeneration can take place rapidly.

The essential steps for achieving people's involvement in resource development and management according to Jimmie Yen–a mass educator in China - are to go to people, live with people, start with what they know and build on what they have.[18]

Van (Forest) Panchayats in Uttar Pradesh Hills

Forest Grievances Committee for Kumaon in 1921 reclassified the forest into two categories as reserved and dereserved forests and recommended the formation of Van Panchayats for the management of the dereserved forest lands. Following this recommendation, till 1985, a total of 4058 Van Panchayats were formed who managed 4,69,326 hectares of forest spread over five districts in the hills of Utter Pradesh. The important objective of this kind of system was to maintain and conserve and possibly improve the productivity of forest and use it (both forest produce and cash proceeds) for the benefit of the local people.[19]

Participation of people in Van Panchayats forest management at the grassroot level is tremendous. The upkeep (vegetative coverage) of U.P. forest is at least as good as in the reserved forest under the Forest Department (FD)[20] and distribution of the forest produce is fairly equitable.

Sukhomajri Village in Himalayas

Sukhomajri is a village in the foothills of the Siwalik range of Himalayas. For several years, this land area had been affected by severe soil erosion, forest denudation and declining land productivity. All the efforts of the government were unsuccessful because of lack of people's participation. But when people organized themselves and took steps, the whole land was reclaimed and it was a high success afterwards.[21]

West Bengal Experience

West Bengal Forest Department is rehabilitating the sal forests through community participation and it agreed to give the villages

25 per cent of all revenues generated from the sale of firewood and timber. This joint venture created a lot of job opportunities, forester-villager relationships improved and the natural forests recovered rapidly. The FD has created nearly 1300 Forest Protection Committees (FPCs). This management system increased the productivity of a number of non-wood forest products.[22]

The Western Semi-Arid Zone

This area includes the dry lands of Western India extending through parts of Gujarat, Rajasthan and Haryana. The people here are semi-nomadic herders and grazers. Their major income is from animal husbandry. So sophisticated indigenous systems of resource management have been developed by the local community to maximize the production of leaf and grass fodder. These system include the spatial, temporal regulation of grazing and fodder collection.[23]

Resource Management in the Himalayas

In 1977, a National Forestry Plan was introduced allowing community in managing the national forest in Nepal, by which the village administrative units were authorized to take over the management responsibility of forests. Thus the local users often exploit forest resources on a rotational basis, allowing them to recover.[24] All these experiences prove that community participation generally is important to the long-term success of local resource management system.[25]

Following these concrete experiences and examples, there can be forest protection committees formed for each of the tribal village or for cluster of tribal villages based on the type of NTFPs collected and the proximity of settlements. The tasks of this committee should be fourfold:

1. To protect the forest and conserve, manage and develop its resources
2. To share the revenue from the forests equally among all the members
3. To run the co-operatives for the NTFPs
4. To choose the beneficiaries for any governmental and non governmental development programmes and projects

All the tribals in a village or cluster of villages should be members in this and the management should be done by a managing committee comprising of duely and democratically elected villagers, a few dedicated forest department officials and a few members from the local non governmental organizations who are involved in tribal development. The reserve forest to be managed by each of this committee should be clearly demarcated. Well defined rules and regulations are indispensable for the smooth management of these committees. These rules and regulations should be formulated after consulting and discussing the local people. Details like the objectives and strategies for achieving this end can be worked out in each village after the formation of Forest Protection Committees. Running a co-operative for NTFPs should be one of the main functions of this committee. Running a co-operative for NTFPs should be one of the main functions of this committee.

The main function of the cooperative is to look after the buying and marketing of the NTFPs collected. Certain amount of revenue in these co-operatives can be set apart for the educational purposes of the children in that particular village. Thus, conservation of biodiversity and development of the tribals, sustainable culling and lucrative marketing should be the focus of these co-operatives.

In spite of the fact that there are many governmental welfare programmes, none of these programmes reach the real needy completely. In some cases, through manipulation by influential persons, rich beneficiaries are selected and in several cases the full amount sanctioned as loan (with subsidy) does not reach the actual beneficiaries. That is why, all the governmental as well as non governmental welfare programmes should go through the forest protection committees. This will facilitate not only seletion of correct and needy beneficiaries but also ensure that the amount sanctioned reach these people without any malpractice.

Finally potential of developing some modern drugs by making use of the traditional wisdom of these Palliyans and Pulayans can be explored. One concrete example for this venture is the development of a drug called 'Jeevani'. It is a restorative drug used for building up stamina. It is made out of *Trichopus zeylanicus* var. *trivancoricus*. The knowledge about the value and use of this drug has been acquired from the Kani tribe living in the South Western Ghats. This value added drug was developed by the scientists of Tropical Botanical

Garden Research Institute (TBGRI) in Trivandrum in Kerala and marketed by Arya Vaidya Pharmacy in Coimbatore.

The tribals who are one of the earliest known occupants of the Palni hills, have been economically exploited and socially subjugated by the other caste people who came from the plains and occupied these hills in the 18th century. Through all the exploitation and subjugation, the tribals have been pushed to the brink of poverty and misery, yet they struggle to eke out a living by exploiting the forest resources. So any further study on the same should aim at the upliftment of the tribals through the enhancement and protection of the forest resources.

References

1. Panchal and Ashwin. 2002. 'Red hot green remedies' in The New Indian Express, September 2, 2002. p. 12.
2. Ravikumar and Madhavi. 2002. Red hot green remedies. In: 'The New Indian Express'. September 2, 2002. p. 12.
3. Sharma and Swati 2002. 'Easy on the body' in the article titled 'Red hot green remedies' in 'The New Indian Express'. September, 2, 2002. p. 12.
4. Vaidya Bhagawan Dash. 1989. Fundamentals of Ayurvedic Medicine. Konark Publishers Pvt. Ltd. Delhi. p. 228.
5. Abdul Kareem, M., Siddarameshwara Hiremath, Nimbalkar, S.D. and Suresh, D.R. 1999. Ethnomedicinal parks–Prefectures of conservation in Amruth pp. 7-20.
6. Burman, B.K. 1989. Problems and Prospects of Tribal Development in North-East India. Economic Political Weekly, April 1, 1989, pp. 693-697.
7. Gadgil, M. 1982. Towards an Indian Conservation Strategy. Paper presented at the Workshop on a New Forest Policy. Indian Social Institute, Delhi, April 12-14.
8. Forests and People. 1988. Resolution adapted by the Workshop on People, Forests and Environment, held on 12th June 1988 at Dahanu, Organised by POWV, Bombay.
9. Chambers, R., Saxena, N.C. and Shah, T. 1989. To the Hands of the Poor, Water and Trees. Oxford and IBH Publishing Co. Pvt. Ltd., New Delhi, pp. 273.

10. Baxi and Upendra, 1983. Towards a Design of Countervailing People's Power in Law and Administration: Agendum for Democratic Law-making. In: Towards a New Forest Policy, (eds.) Fernandes, W. and Kulkarni, S. Indian Social Institute, New Delhi, pp. 111-133.
11. Gadgil, M. 1983. Forestry with a Social purpose. In: Towards a New Forest Policy, (eds.) Fernandes, W. and Kulkarni, S. Indian Social Institute, New Delhi, pp. 111-133.
12. Bisht, B.S. 1994. Tribes of India, Nepal, Tibet Borderland–A Study of Cultural Transformation. GIAN Publishing House, New Delhi–110 002. pp. 285.
13. Berdoloi, B.N. 1989. 'Strategies for Tribal Development in North-East India with Special Reference to Assam'. In: Raha, M.K., and Palash Chandra Coomar (ed.) Tribal India–Problem Development Project. GIAN Publishing House, New Delhi–110 002. p. 115-116.
14. Tete and Peter. 1986. A Missionary Social Worker in India, Fr. J.B. Hoffmann. The Chota Nagpur Tenancy Act and the Catholic Cooperatives. Satya Bharathi, Ranchi, pp. 191.
15. Vyasula and Vinod. 1981. Alternative Development strategies for Koraput. Paper presented at the Seminar on Alternative Development Strategies: Their Relevance to India (Lucknow, Feb. 21-22, 1981).
16. FAO. 1982. As reported in participation in forestry. Paper by Siteke Kuperus presented at a workshop. October 1987. Department of Forest Management, Agricultural University, Wageningen.
17. Agarwal, Anil and Sunita, N. 1990. Green Villages: A Strategy for Environmentally Sound and Participatory Rural Development. Centre for Science and Environment, pp. 52.
18. Melkania, N.P. 1990. Forest Management–People's Participation. Proceedings of the MAB Regional Training Workshop of Tropical Forest Ecosystem Conservation and Development in South and South East Asia. Kerala Forest Research Institute. Peechi, Kerala, pp. 149-157.
19. Ballabh, Vishwa and Singh, K. 1987. People's Participation in Management of Natural Forests–the Experience of Lessons from

Van Panchayats of Utter Pradesh Hills (mimeo). Institute of Rural Management, Anand.

20. Gadgil, M. 1982. Towards an Indian Conservation Strategy. Paper presented at the Workshop on a New Forest Policy. Indian Social Institute, Delhi, April 12-14.

    Guha, R. 1983. Forestry in British and Post-British India. Economic Political Weekly, October 29, VII. 18, pp. 45-46.

    Saxena, N.C. 1987. Commons, Trees and the Poor in Uttar Pradesh Hills. ODI. Social Forestry Network Publication, pp. 43.

21. Chopra, K., Gopal, K. and Murthi, M.M. People's Participation and common Property Resources. Economic Political Weekly, December 23-30, 1989, pp. A-189–A-195.

22. Poffenberger and Mark. 1990. Joint Management of Forest Lands: Experiences From South Asia. A Ford Foundation Programme Statement, pp. 49.

23. Jodha, N.S. 1988. Fuel and Fodder Management System in the Arid Region of Western Rajasthan, Kathmandu. International Centre for Integrated Mountain Development.

24. Poffenberger and Mark. 1990. Joint Management of Forest Lands: Experiences From South Asia. A Ford Foundation Programme Statement, pp. 49.

25. Korten, D.C. 1986. Community Management: Asian Experiences and Perspective. New Haven, C.T. Kumarian Press.

Appendices

Appendix 1
RAINFALL DATA FOR KODAIKANAL

Year	Jan.	Feb.	Mar.	Apr.	May	June	July	Aug.	Sept.	Oct.	Nov.	Dec.	Total
1956	01.84	00.10	00.70	05.67	01.41	06.46	01.01	04.12	05.12	16.66	13.76	02.38	59.23
1957	–	01.11	01.25	03.95	04.87	03.91	03.12	04.82	03.66	13.20	13.28	00.83	62.00
1958	01.20	00.45	09.40	06.03	09.48	02.04	01.51	04.54	04.15	05.23	07.27	02.07	53.64
1959	01.34	00.68	–	04.68	10.62	06.17	02.98	05.89	10.23	10.64	15.59	05.56	74.33
1960	–	00.51	02.46	16.39	06.32	02.08	10.90	04.57	06.66	12.90	18.83	01.46	83.08
1961	03.45	00.35	00.36	05.14	03.99	03.95	05.47	05.06	04.37	10.08	06.77	08.49	57.48
1962	01.10	04.74	03.28	07.09	06.00	01.28	03.56	05.57	06.74	14.20	03.07	04.68	61.31
1963	08.38	01.47	03.09	03.38	03.51	04.44	05.07	04.55	08.23	13.69	12.78	09.80	77.39
1964	00.13	–	02.49	03.63	07.30	03.58	08.52	04.11	04.85	12.82	04.78	09.78	61.99
1965	00.19	00.97	02.53	05.74	04.77	01.71	03.45	08.79	04.58	15.61	03.81	11.81	53.96
1966	01.65	–	02.99	01.83	01.23	06.40	08.35	06.73	08.74	10.85	11.62	06.03	66.42
1967	00.50	00.05	02.15	03.55	02.76	06.52	04.36	05.84	04.08	17.28	05.09	04.76	56.94
1968	00.59	02.04	05.21	07.11	04.64	04.93	04.03	03.80	10.98	07.69	05.50	05.49	62.01
1969	00.85	01.03	00.88	05.90	02.66	01.58	02.53	11.10	01.76	11.99	08.98	08.44	57.75
1970	02.46	01.46	12.30	10.45	00.80	04.15	05.54	05.66	10.35	07.79	01.72	01.72	63.50
1971	01.81	04.53	01.83	04.43	13.59	01.17	09.29	08.47	08.49	12.27	03.13	14.02	83.03

Contd...

Appendix 1–Contd...

Year	*Jan.*	*Feb.*	*Mar.*	*Apr.*	*May*	*June*	*July*	*Aug.*	*Sept.*	*Oct.*	*Nov.*	*Dec.*	*Total*
1972	00.23	00.13	–	01.86	12.80	02.51	03.70	03.11	13.54	12.03	08.63	05.44	63.99
1973	06.93	08.75	06.71	04.79	05.86	05.79	06.93	08.75	08.58	05.79	05.50	02.89	68.46
1974	–	00.45	00.78	05.08	05.62	02.62	07.64	04.30	14.09	06.39	00.79	02.75	50.51
1975	01.16	01.08	04.81	06.47	06.14	00.80	08.46	04.35	11.02	04.13	06.18	02.45	57.05
1976	–	–	02.46	03.92	02.81	06.67	04.75	05.75	06.68	06.89	11.87	03.86	55.06
1977	00.03	02.89	04.54	09.61	07.57	02.51	04.57	09.27	06.95	24.14	20.82	00.45	93.35
1978	00.09	02.28	00.17	04.92	05.57	01.26	02.69	08.68	08.08	02.08	12.18	14.22	62.48
1979	00.08	09.17	00.82	02.92	04.51	04.56	07.17	03.66	13.56	11.87	34.25	01.47	94.04
1980	–	–	04.51	08.13	03.68	01.97	03.28	04.96	07.14	10.13	10.95	02.18	56.93
1981	01.20	00.17	04.03	03.46	06.22	04.82	06.05	05.01	13.21	09.53	04.69	04.72	63.11
1982	00.15	–	00.68	03.15	05.79	01.26	04.80	00.77	09.87	15.57	07.50	03.49	52.93
1983	00.33	00.55	01.36	02.41	06.29	05.32	05.35	05.80	08.12	08.12	06.00	12.72	60.36
1984	04.68	05.94	01.22	07.98	01.70	01.25	06.08	03.03	11.18	10.24	06.18	05.39	79.87
1985	04.99	–	01.71	11.29	03.88	02.21	01.78	02.63	09.08	06.10	07.90	04.27	57.84
1986	01.78	04.43	01.02	01.20	04.00	02.59	02.30	01.25	08.81	05.75	06.51	02.49	42.13
1987	00.88	00.03	03.88	00.19	04.36	04.78	01.92	04.96	14.82	15.04	06.70	17.06	74.62
1988	–	00.06	03.74	12.38	00.65	02.06	06.30	06.44	07.84	02.35	04.27	01.60	49.71
1989	00.05	–	05.76	06.90	05.48	02.90	06.09	02.61	14.03	07.27	08.87	02.17	62.13

Source: Kodaikanal Boat Club Centenary Commemoration Volume, 1990.

Appendix 2A
CLAN DETAILS AND THEIR DISTRIBUTION

Sl.No.	*Name of Clan*	*Households*		*Population*	
		Total	*%*	*Total*	*%*
1.	Palian	44	6.40	156	6.42
2.	Azhagarasan	17	2.47	71	2.92
3.	Lengan	64	9.30	228	9.38
4.	Vembilian	51	7.41	166	6.83
5.	Kullan	15	2.18	48	1.98
6.	Kudian	81	11.77	285	11.73
7.	Peruman	34	4.94	128	5.27
8.	Odaiyan	13	1.89	48	1.98
9.	Kosavan	20	2.91	65	2.67
10.	Narayan	25	3.63	98	4.03
11.	Kovan	50	7.27	192	7.90
12.	Chiruman	44	6.40	152	6.26
13.	Manigan	9	1.31	36	1.48
14.	Rettaian	22	3.20	84	3.46
15.	Mottaian	34	4.94	115	4.73
16.	Chinandi	16	2.33	53	2.14
17.	Sadaiyandi	37	5.38	121	4.28
18.	Chamanan	7	1.02	27	1.11
19.	Chinnanchi	2	0.29	8	0.33
20.	Paliya Kuppan	3	0.44	11	0.45
21.	Adiyank Kannan	8	1.16	30	1.23
22.	Kudumpi	4	0.58	10	0.41
23.	Mallan	15	2.18	54	2.22
24.	Vadaraiyan	11	1.60	39	1.60
25.	Chenganan	14	2.03	43	1.77
26.	Udaiyandi	2	0.29	5	0.21
27.	Karunk Chenganam	1	0.15	2	0.08
28.	Maavati	3	0.44	9	0.37
29.	Unspecified	42	6.10	147	6.05

Appendix 2B
DEITIES WORSHIPPED

Sl.No.	*Deity*	*Households*		*Population*	
		Total	*%*	*Total*	*%*
1.	Palichi Macca	43	6.25	152	6.26
2.	Agarathammal	15	2.18	61	2.51
3.	Meenachi Rengarsami	66	9.59	249	10.25
4.	Nagarsami	52	7.56	172	7.08
5.	Senganatchi	1	0.15	2	0.08
6.	Kudiamalai	70	10.17	237	9.75
7.	Ramarsami	29	4.22	111	4.57
8.	Peria Pattarai	5	0.73	23	0.95
9.	Karuvarai Andi	9	1.31	20	0.82
10.	Muthayi Pattayi	13	1.89	51	2.10
11.	Kurumuchi Atha	37	5.38	141	5.81
12.	Udharakaliamma	17	2.47	64	2.63
13.	Thangamar Atha	26	3.78	95	3.91
14.	Chokkan	24	3.49	67	2.76
15.	Kumara Kandan	27	3.92	92	3.79
16.	Pudhanachi Sadayandi	17	2.47	61	2.51
17.	Mangalanathan	44	6.40	161	6.63
18.	Andiappar	13	1.89	56	2.30
19.	Malaichi Macca	1	0.15	3	0.12
20.	Periana Sami	17	2.47	49	2.02
21.	Periapattali Thatha	4	0.58	14	0.58
22.	Kulachi Natchiammal	6	0.87	15	0.62
23.	Karuppana Sami	18	2.62	58	2.39
24.	Madurai Veeran	7	1.02	23	0.95
25.	Kaliamman	11	1.60	31	1.28
26.	Pannirendu Deivam	1	0.15	4	0.16
27.	Pavanachi	2	0.29	7	0.29
28.	Kariasami	4	0.58	16	0.66

Contd...

Appendix 2B–Contd...

Sl.No:	Deity	Households		Population	
		Total	%	Total	%
29.	Thotta Malayandi	11	1.60	36	1.48
30.	Arakalandi	1	0.15	3	0.12
31.	Madhavar	1	0.15	3	0.12
32.	Karahanatchi	1	0.15	4	0.16
33.	Mottayandi	1	0.15	4	0.16
34.	Appuchi Sami	3	0.44	10	0.41
35.	Vedimallan	2	0.29	9	0.37
36.	Kunrunathan	2	0.29	5	0.21
37.	Mayautharam	13	1.89	45	1.85
38.	Poovandichi Amman	21	3.05	74	3.05
39.	Vettai Karan Sami	6	0.87	26	1.07
40.	Karukattamma	6	0.87	21	0.86
41.	Sellandi	3	0.44	13	0.53
42.	Solachiamma	1	0.15	4	0.16
43.	Kovi Kizhavi	1	0.15	2	0.08
44.	Sangili Karuppan	2	0.29	4	0.16
45.	Seelakari	3	0.44	17	0.70
46.	Neelinachi	3	0.44	17	0.70
47.	Peria Solayan	7	1.02	30	1.23
48.	Peria Andavar	2	0.29	6	0.25
49.	Kal Appuchi	2	0.29	7	0.29
50.	Neela Malayandi	6	0.87	18	0.74
51.	Perunattu Makkal	2	0.29	10	0.41
52.	Velangappar	1	0.15	3	0.12
53.	Unspecified	10	1.45	37	1.52
	TOTAL	**688**		**2430**	

Appendix 2C
PRINCIPAL OCCUPATION OF THE PULAYANS

Sex	Unemployed		Daily Wage		MFP Collection		Transportation by Horse		Cattle Grazing		Government Job		Agriculture	
	Total	%	Total	%	Total	%	Total	%	Total	%	Total	%	Total	%
Male	428	17.61	820	33.47	2	0.08	2	0.08	3	0.12	9	0.37	6	0.25
Female	391	16.09	765	31.48	0	0.00	0	0.00	0	0.00	0	0.00	4	0.16
Total	819	33.70	1585	65.23	2	0.08	2	0.08	3	0.12	9	0.37	10	0.41

Appendix 2D
AGE WISE LITERACY LEVEL OF THE PULAYANS

Age Group	Illiterate	Literacy Level											
		I	II	III	IV	V	VI	VII	VIII	IX	X	XI	XII	Total
0–5	275	12	1	0	0	0	0	0	0	0	0	0	0	289
6–10	180	34	45	23	11	13	2	0	0	0	0	0	0	308
11–15	105	8	6	21	11	16	14	8	7	4	0	0	0	199
16–20	202	2	6	7	10	26	11	10	13	4	2	1	2	296
21–25	204	2	1	11	9	23	18	11	17	8	4	0	1	309
26–30	233	3	2	14	5	31	9	6	17	9	4	0	1	334
31–35	121	1	4	4	8	29	7	9	14	2	2	0	0	201
36–40	126	1	3	8	3	19	13	11	8	3	0	0	0	195
41–45	90	0	3	1	0	5	5	3	5	1	0	0	0	113
46–50	76	0	0	0	1	8	3	2	3	0	0	0	0	93
51–55	34	0	0	0	0	1	0	0	0	0	0	0	0	35
56–60	30	0	0	0	0	1	0	0	0	0	0	0	0	31
61–65	10	0	0	0	0	0	0	0	0	0	0	0	0	10
66–70	10	0	1	0	0	0	0	0	0	0	0	0	0	11
Above 70	5	0	0	0	0	1	0	0	0	0	0	0	0	6
	Total	63	72	89	58	173	82	60	84	31	12	1	4	2430

Appendix 2E
SEXWISE LITERACY LEVEL FOR THE PULAYAN VILLAGES

Education	*Male*	*%*	*Female*	*%*	*Total*
Uneducated	738	30.37	963	39.63	1701
I	34	1.40	29	1.19	63
II	45	1.85	27	1.11	72
III	61	2.51	28	1.15	89
IV	38	1.56	20	0.82	58
V	136	5.60	37	1.52	173
VI	65	2.67	17	0.70	82
VII	50	2.06	10	0.41	60
VIII	69	2.84	15	0.62	84
IX	21	0.86	10	0.41	31
X	9	0.37	3	0.12	12
XI	0	0.00	1	0.04	1
XII	4	0.16	0	0.00	4
Total	1270	52.26	1160	47.74	2430

Appendix 2F
LAND OWNERSHIP PATTERN

Number of Acres Owned	*Nature of Lands Owned (Number of Households)*		
	Patta	*Porampokku*	*Bhimo*
1	49	49	14
2	32	48	14
3	3	31	10
4	–	4	1
5	2	2	–
Total	132	204	82
Mean/HHs	1.53	1.97	1.95

Appendix 3A
AGE DISTRIBUTION VERSUS TOTAL POPULATION

Sl.No.	*Age Group*	*Male*	*%*	*Female*	*%*	*Total*	*%*
1.	0–5	60	6.05	47	4.74	107	10.79
2.	6–10	80	8.06	66	6.65	146	14.72
3.	11–15	39	3.93	48	4.84	87	8.77
4.	16–20	57	5.75	61	6.15	118	11.90
5.	21–25	62	6.25	88	8.87	150	15.12
6.	26–30	70	7.06	48	4.84	118	11.90
7.	31–35	35	3.53	29	2.92	64	6.45
8.	36–40	27	2.72	30	3.02	57	5.75
9.	41–45	24	2.42	21	2.19	45	4.54
10.	46–50	44	4.44	24	2.42	68	6.85
11.	51–55	9	0.91	5	0.50	14	1.41
12.	56–60	9	0.91	4	0.40	13	1.31
13.	61–65	1	0.10	1	0.10	2	0.20
14.	66–70	1	0.10	1	0.10	2	0.20
15.	> 70	1	0.10	0	0.00	1	0.10
	Total	519	52.32	473	47.68	992	100.00

Appendix 3B
DEITIES WORSHIPPED

Sl.No.	*Deities*	*Households*		*Population*	
		Total	*%*	*Total*	*%*
1.	Palichi Macca	111	40.07	395	39.82
2.	Kurumuchi Atha	1	00.36	3	00.30
3.	Madurai Veeran	2	00.72	8	00.81
4.	Vandi Kaliamman	32	11.55	110	11.09
5.	Kaliamman	66	23.83	230	23.19
6.	Mariamman	42	15.16	164	16.53
7.	Pannirendu Deivan	16	5.78	58	5.85
8.	Veera Kamachi	1	00.36	6	00.60
9.	Unspecified	6	02.17	18	01.81
	Total	**277**	**100.00**	**992**	**100.00**

Appendix 3C
SEXWISE LITERACY LEVEL FOR THE PALIYAN VILLAGES

Education	*Male*	*%*	*Female*	*%*	*Total*
Uneducated	469	47.28	435	43.85	904
I	13	1.31	9	0.91	22
II	8	0.81	7	0.71	15
III	5	0.50	8	0.81	13
IV	2	0.20	2	0.20	4
V	9	0.91	5	0.50	14
VI	1	0.10	3	0.30	4
VII	6	0.60	1	0.10	7
VIII	3	0.30	2	0.20	5
IX	2	0.20	1	0.10	3
X	1	0.10	0	0.00	1
Total	519	52.32	473	47.68	992

Appendix 3D
AGEWISE LITERACY LEVEL OF THE PALIYAN VILLAGES

Age Group	*Illiterate*	*Literacy Level*										
		I	*II*	*III*	*IV*	*V*	*VI*	*VII*	*VIII*	*IX*	*X*	*Total*
0–5	103	4	0	0	0	0	0	0	0	0	0	0
6–10	112	16	10	4	2	1	1	0	0	0	0	146
11–15	67	0	4	7	1	4	2	2	0	0	0	87
16–20	101	0	1	1	1	4	1	4	2	2	1	118
21–25	144	0	0	0	0	2	0	1	2	1	0	150
26–30	114	0	0	1	0	2	0	0	1	0	0	118
31–35	63	1	0	0	0	0	0	0	0	0	0	64
36–40	56	1	0	0	0	0	0	0	0	0	0	57
41–45	44	0	0	0	0	1	0	0	0	0	0	45
46–50	68	0	0	0	0	0	0	0	0	0	0	58
51–55	14	0	0	0	0	0	0	0	0	0	0	14
56–60	13	0	0	0	0	0	0	0	0	0	0	13
61–65	2	0	0	0	0	0	0	0	0	0	0	2
66–70	2	0	0	0	0	0	0	0	0	0	0	2
Above 70	1	0	0	0	0	0	0	0	0	0	0	1
Total	904	22	15	13	4	14	4	7	5	3	1	992

Appendix 3E
LAND OWNERSHIP PATTERN

Number of Acres Owned	*Nature of Lands Owned (Number of Households)*		
	Patta	*Porampokku*	*Bhimo*
1.	5	2	6
2.	6	4	6
3.	1	3	5
4.	–	–	–
5.	–	–	–
Total land owned	20	19	33
Mean/HHs	1.67	2.11	1.94

Appendix 4
THE RESERVE FORESTS (RF.) IN KODAIKANAL FOREST DIVISION

SL.No.	*Name of RF.*	*Extent in Ha.*	*%*
1.	UNJALAACHI	208.82	0.51
2.	PALNI HILLS SOUTHERN SLOP	3254.62	7.99
3.	KEELANNAVAYAL	226.37	0.56
4.	GLEN FALLS	14.76	0.04
5.	GUNDAR VALLEY	1809.04	4.44
6.	SAMIKANAL	655.66	1.61
7.	POST OFFICE PLANTATION	0.40	0.001
8.	TIGER SHOLA	506.74	1.24
9.	MULAIYAR	959.94	2.36
10.	ANGIRISHOLA	30.82	0.08
11.	GUNDAR VALLEY EXTENSION	2229.39	5.47
12.	AMPT HILL DOWN	12791.00	31.40
13.	KOOKAL	346.01	0.85
14.	PERUMAL MALAI SHOLA	113.10	0.28
15.	PAMBAR	1710.14	4.20
16.	POOMBARAI	6.83	0.02
17.	PERUMALMALAI	2598.04	6.38
18.	ARUNKANAL	493.51	1.21
19.	SENGALVARAYAR	1821.35	4.47
20.	BOMBAY SHOLA	8.32	0.02
21.	KARUNGAL THONIMEDU	1134.47	2.78
22.	AMBURUVI	3543.86	8.70
23.	ADUKKAM	2277.29	5.59
24.	AYAKUDI ZAMIN R.L.	4000.00	9.82
	DIVISION TOTAL	**40740.48**	

Index

A

Abutilon indicum 138
Acacia caesia 139
Acacia canescens 139
Acacia pennata 139
Acacia sinuata 139
Acalypha fruticosa 139
Acalypha indica 87, 139
Acalypha racemosa 139
Acculturation 209
Acorus calamus 140
Adhatoda zeylanica 87, 140
Adiantum raddianum 140
Advance and Loan Systems 62
Aerva lanata 140
Agaricus campestris 140
Agave cantala 140
Ageratum conyzoides 141
Ailanthus excelsa 141
Albizia amara 141
Albizia chinensis 141
Aleurites moluccana 141
Allium cepa 141
Allium sativum 141
Aloe vera 142
Alphonsea sclerocarpa 142
Alpinia calcarata 142
Alternanthera pungens 142
Alternanthera sessilis 142
Alternanthera viridis 142
Alysicarpus vaginalis 142
Amaranthus caudatus 142
Amaranthus spinosus 143
Amaranthus tricolor 143

Amaranthus viridis 143
Ananas comosus 143
Anaphalis lawii 143
Animal hunted 111
Animism 77
Anisochilus carnosus 143
Anogeissus latifolia 143
Apatains 118
Appalling conditions 199
Aristida setacea 143
Aristolochia tagala 143
Artemisia nilagirica 144
Artocarpus heterophyllus 144
Arundo donax 144
Asparagus racemosus 144
Asteraceae 96
Asystasia gangetica 144
Australoid 50
Azadirachta indica 87, 144

B

Baeolepis nervosa 145
Baigas 118
Balloon vine or blister creeper 135
Bambusa bambos 145
Barbula convoluta 145
Basella alba 145
Begonia floccifera 145
Begonia malabarica 145
Benincasa hispida 145
Bentinkia condapanna 146
Beta vulgaris 146
Bhotias 117
Bidens pilosa 146
Black Dammar 133
Black magic 98
Blepharis maderaspatensis 146
Blumea mollis 146
Blumea oxyodonta 146
Blumea virens 146
Boerhavia diffusa 147
Botanical gardens 208
Brassica juncea 147
Brassica nigra 147
Brassica oleracea 147
Buddleja asiatica 147

C

Cajanus cajan 147
Calamus rotang 147
Calotropis procera 148
Canarium strictum 148
Cansjera rheedii 148
Canthium dicoccum 148
Capsicum frutescens 148
Cardiospermum canescens 148
Cardiospermum halicacabum 149
Carica papaya 149
Carissa carandas 134, 149
Caryota urens (Indian Sago Jaggery) 137, 149
Cassia leschenaultiana 149
Cassia sophera 149
Casuarina 82
Casuarina litorea 150

Celebration of the Village Festivals 65
Celebrations, Rits and Rituals of the Tribals 65
Centella asiatica 150
Chamaecrista absus 150
Changes 201
Changing situation 198
Chenopodium ambrosoides 150
Child Birth 69
Chrysanthemum indicum 150
Chumash Indians 115
Cicer arietinum 150
Cinnamomum macrocarpum 151
Cinnamomum zeylanicum 151
Cipadessa baccifera 151
Cissampelos pariera 151
Cissus quadrangularis 151
Citrus aurantium 151
Citrus limon 151
Citrus pennivesiculata 152
Clans and Deities 51
Clausena heptaphylla 152
Clausena indica 152
Clematis munroiana 152
Cleome gynandra 152
Cleome monophylla 152
Clerodendrum serratum 153
Climbers 97
Co-operative 220
Coccinia grandis 153
Cocos nucifera 153
Coffea arabica 153
Coffea robusta 153
Coffee Cultivation 108
Colocasia esculenta 153
Commelina benghalensis 153
Commelina hirsuta 154
Commercialization 198
Commiphora caudata 73, 154
Common diseases 77
Common Property Resources 220
Conclusions 206
Conservation 214
Constitutional Safeguards 216
Cooking 111
Cooking and Food Habits 62
Corchorus aestuans 154
Cordia obliqua 154
Coriandrum sativum 154
Crassocephalum crepidioides 154
Crotalaria pallida 154
Croton bonplandianus 154
Cryptolepis buchanani 155
Ctenolepis garcinii 155
Cultivated crops 38
Cultivated plants 108
Cultural practices 51
Cuminum cyminum 155
Curative methods 80
Curcuma longa 155
Cyathula prostrata 155
Cymbopogon flexuosus 156
Cymbopogon martinii 156
Cymbopogon nardus 156

Cymbopogon travancorensis 156
Cynodon dactylon 156
Cynoglossum zeylanicum 155
Cyphomandra betacea 156

D

Dance 66
Dances and Music 75
Date Palm 132
Datura arborea 156
Datura innoxia 157
Datura metel 157
Daucus carota 157
Decentralised knowledge 207
Deforestation 5
Deities worshipped 51
Department for the Adi-Dravidar 218
Desmodium triflorum 157
Development of the Tribals 218
Development projects 197
Dicots 97
Didymocarpus innominatus 157
Dioscorea alata 157
Dioscorea bulbifera 157
Dioscorea esculenta 158
Dioscorea pentaphylla 158
Diospyros nigricans 158
Diplocyclos palmatus 158
Disappearing Traditional Lifestyle 199
Diseases 97
Dodonaea viscosa 158
Dolmen circles 48
Dolmens 48
Domestic needs 110
Drawbacks of Ethnomedicine 209
Dress habits 61
Dumasia villosa 158

E

Eclipta prostrata 158
Economic botany 11, 12
Economic importance 126
Economic policies 198
Education 52
Elettaria cardamomum 137, 159
Eleusine coracana 159
Emilia sonchifolia 159
Entada pursaetha 159
Erythrina variegata 159
Ethnobotany 10
Ethnobotany as an Emerging Science 14
Ethnobotany in India 12
Ethnoecology 10
Ethnomedicinal park 214
Ethnomedicine 78
Eucalyptus globulus 159
Eupatorium adenophorum 159
Euphorbia hirta 159
Euphorbia rosea 160
Euphorbia thymifolia 160
Euphorbia tirucalli 73, 160
Evil eyes 98

Evolvulus alsinoides 160
Exotic flora 38

F

Fabaceae 96
Fast Moving Herbal Goods 212
Ferula asafoetida 160
Festival celebrations 65
Ficus benjamina 160
Ficus callosa 160
Ficus glomerata 73
Ficus hispida 160
Ficus racemosa 161
Ficus religiosa 73, 161
Ficus tsjahela 161
Firewood 111
Flemingia macrophylla 161
Floristic Wealth of the Western Ghats 27
Foeniculum vulgare 161
Folk Tales among these Tribals 62
Folklore 11
Food 107
Food Habits 61
Forest policies 196
Forest Policy 216
Funeral Rites 74
Funeria hygrometrica 161

G

Gallnut 126
Geographical fragmentation 198
Gingeli seeds 74
Girardinia diversifolia 162
Gmelina arborea 75, 162
Gmelina asiatica 162
God man 65
Gooseberry 131
Grasses and leaf fodder 119
Grevillea robusta 162
Grewia tiliifolia 162
Gymnema sylvestre 162
Gymnosperm 97

H

Health Status of the Tribals 76
Helicteres isora 163
Hemidesmus indicus 87, 163
Hemionitis arifolia 163
Heracleum rigens 163
Herbal healthcare 212
Herbalists 81
Herbs 97
Hibiscus cannabinus 163
Hibiscus rosa-sinensis 163
Himalayas 222
Homogenization 198
Honey 138
Housing and Land Ownership 52
Human life and culture 1
Hydrocotyle javanica 163

I

Impatiens balsamina 164

Inadequacy of the 'English medicines' 213
Indian Kino Tree 135
Indu Kodi 132
International Convention 215
Ipomoea obscura 164
Ipomoea quamoclit 164
Irulas 117
Ixora nigricans 164

J

Jasminum cuspidatum 164
Jeevani 223
Justicia procumbens 164

K

Kalanchoe bhidei 165
Kalanchoe grandiflora 165
Kattu Milagu 132
Khasi and Hainatia hills 116
Kleinia grandiflora 165
Kodaikanal 29
Kotas 117

L

Lablab purpureus 165
Lagenaria siceraria 165
Lantana camara 165
Lawsonia inermis 165
Lemon Grass 135
Lepisanthes tetraphylla 136, 166
Leptadenia reticulata 166
Leucas aspera 166
Leucas cephalotes 166
Leucas indica 167
Leucas vestita 167
Lifestyle 200
Lindernia oppositifolia 167
Local health traditions 208
Lok Swasthya Parampara Samvardhan Samithi 212
Lycianthes laevis 167
Lycopersicon esculentum 167

M

Mahonia leschenaultii 167
Malai Shenbagam 133
Malvastrum coromandelianum 167
Mangifera indica 168
Marriage 72
Medicinal herbs 13
Medicinal plants 15, 97
Medicinemen 41, 209
Melia azedarach 168
Mentha arvensis 168
Michelia nilagrica 134, 168
Mimosa pudica 168
Miris 118
Miris or Mishings 117
Miskutoes 116
Modern Medicine Versus Traditional Medicine 209
Monocots 97
Moringa pterygosperma 168
Moss 126
Mukia leiosperma 168

Mukia maderaspatana 169
Murraya koenigii 169
Musa paradisiaca 169
Music 66
Mustard seeds 74
Myristica fragrans 169

N

Naravelia zeylanica 169
Nationalization 122
Networking 208
NGO 219
Nishis 118
Nomadic tribes 49
Non-Timber Forest Products 16, 112

O

Occupation 51
Ocimum americanum 169
Ocimum gratissimum 170
Ocimum tenuiflorum 87, 170
Opilia amentacea 170
Orthosiphon aristatus 170
Oryza sativa 171
Osyris arborea 171
Other methods 97
Oxalis corniculata 171
Oxalis latifolia 171

P

Palliyans 4, 53
Palni Hills 28
Panchasheel 217
Papaver somniferum 171
Participant observation 39
Passiflora calcarata 136, 171
Passiflora edulis 172
Passion Fruit 133
Pauri Bhuirya 118
PEAK 219
Pennisetum typhoides 172
Pentatropis capensis 172
People of Kwara state 116
People's Participation 220
Pergularia daemia 172
Phaseolus calcaratus 172
Phoenix loureirii 172
Phoenix sylvestris 172
Phyllanthus amarus 172
Phyllanthus debilis 173
Phyllanthus emblica 173
Physical features 60
Phytomedicine 78
Pinus roxburghii 173
Piper betle 173
Piper mullesua 135, 173
Piper nigrum 174
Piper trioicum 174
Plants used for culture 98
Plectranthus amboinicus 174
Plectranthus barbatus 174
Plectranthus caninus 174
Plectranthus malabaricus 174
Polygala arillata 175
Polygala arvensis 175

Polygonum chinense 175
Polytrichum densifolium 175
Pongamia pinnata 175
Portulaca oleracea 175
Portulaca pilosa 176
Portulaca quadrifida 176
Poverty 197
Preventive 80
Primitive people 1
Professional hindrances 209
Psidium guajava 176
Psilotrichum scleranthum 176
Pteridophytes 97
Pterocarpus marsupium 176
Puberty 70
Pulayan households 51
Pulayans 4
Punica granatum 171

R

Raphanus sativus 176
Recreational Hobbies 60
Religion 81
Revitalisation programme 213
Revival of Ethnomedicine 212
Ricinus communis 176
Rites of Passage 69
Ritualism 81
Ritualists 81
Rituals 69
Rubia cordifolia 177
Rubus niveus 177
Ruta chalepensis 177

S

Sansevieria roxburghiana 87, 177
Santal communities 78
Sapindus emarginatus 134, 178
Scheduled Tribes 2
Schefflera stellata 178
Sechium edule 178
Senecio scandens 178
Senna alexandrina 178
Senna auriculata 178
Senna occidentalis 178
Senna tora 178
Sesamum indicum 179
Sesbania grandiflora 179
Setaria 73
Sholas 37
Shrubs 97
Sida acuta 179
Sida cordata 179
Sida glutinosa 179
Sida rhombifolia 179
Soapnut 133
Socio-Cultural Dimensions of Health 81
Solanum anguivi 180
Solanum erianthum 180
Solanum giganteum 180
Solanum melongena 180
Solanum nigrum 180
Solanum torvum 181
Solanum trilobatum 87
Solanum tuberosum 181
Solanum virginianum 180

Sorghum vulgare 181
Spilanthes calva 181
Stenosiphonium russelianum 181
Stephania hernandifolia 181
Strengths of Ethnomedicine 210
Strobilanthes kunthiana 182
Stylosanthes fruticosa 182
Styrax benzoin 136, 182
Sukhomajri 221
Sullungs 118
Sulungs 118
Supernatural 80
Sustainable development 215
Syzygium aromaticum 182
Syzygium cumini 182

T

Tabernaemontana gamblei 182
Talinum portulacifolium 183
Talisman 82
Tamangs 116
Tamarind 119
Tamarindus indica 183
Tasar silk 119
Tephrosia fusca 183
Tephrosia purpurea 183
Tephrosia tinctoria 183
Terminalia arjuna 134, 183
Terminalia chebula 184
Tetrastigma lanceolarium 184
Tharus 118
Thellikkai 133
Themeda cymbaria 184
Themeda tremula 184
Themeda triandra 184
Thespesia populnea 184
Thunbergia alata 184
Todas 117
Toddalia asiatica 184
Tools 40
Toona ciliata 185
Trachyspermum ammi 185
Trees 97
Tribal development 215
Tribal ethnomedicine 77
Tribal healers 81
Tribal health care 207
Tribal Perception of Etiology, Health and Medicine 79
Tribals and Forests 4
Tribals in India 2
Tribals in Orissa 118
Tribals in Tamil Nadu 3
Trichosanthes lobata 185
Tridax procumbens 185
Trigonella foenum-graecum 185
Triumfetta rotundifolia 185
Typhonium trilobatum 186

V

Van Panchayats 221
Vegetation 35
Vernonia anthelmintica 137, 186
Vigna mungo 186
Vigna unguiculata 186
Vitex negundo 87, 186
Vitex trifolia 186

W

Waltheria indica 187

Wedelia chinensis 187

West Bengal 221

Western Ghats 24

Western Semi-Arid Zone 222

Witchcraft 98

X

Xanthium indicum 187

Y

Yams 61

Z

Zanthoxylum rhetsa 187

Zanthoxylum tetraspermum 187

Zehneria maysorensis 187

Zingiber officinale 188

Ziziphus oenoplia 188

Ziziphus rugosa 188